Astronomers' Observing Guides

Other titles in this series

The Moon and How to Observe It
Peter Grego

Related titles

Field Guide to the Deep Sky Objects
Mike Inglis

Deep Sky Observing
Steven R. Coe

The Deep-Sky Observer's Year
Grant Privett and Paul Parsons

The Practical Astronomer's Deep-Sky Companion
Jess K. Gilmour

Observing the Caldwell Objects
David Ratledge

Choosing and Using a Schmitt-Cassegrain Telescope
Rod Mollise

James Mullaney

Double and Multiple Stars and How to Observe Them

With 45 Figures

Series Editor: Dr. Mike Inglis, FRAS

Cover illustration: A red dwarf, shown in the background, is at the mercy of its highly magnetic white dwarf companion in this depiction of a so-called intermediate polar binary system. Gas lost from the red star is channelled toward the white dwarf's poles along magnetic field lines, where it emits the X-rays that make these systems so powerful.

All illustrations noted 'Courtesy of *Sky & Telescope*' in the captions are copyright *Sky & Telescope* and Sky Publishing Corporation, and are used with their kind permission.

British Library Cataloguing in Publication Data
Mullaney, James
   Double and multiple stars and how to observe them.
   (Astronomers' observing guides)
   1. Double stars-Observers' manuals   2. Double stars-
   Observations   3. Multiple stars-Observers' manuals
   4. Multiple stars-Observations
   I.Title
   523.8 41
ISBN 1852337516

Library of Congress Cataloging-in-Publication Data
Mullaney, James.
   Double and multiple stars and how to observe them / James Mullaney.
      p. cm — (Astronomers' observing guides)
   Includes bibliographical references and index.
   ISBN 1-85233-751-6 (acid-free paper)
    1. Multiple stars–Observations. I. Title. II. Series.

QB821.M83 2005
523.8 41—dc22   2004065359

Apart from any fair dealing for the purposes of research or private study, or criticism or review, as permitted under the Copyright, Designs and Patents Act 1988, this publication may only be reproduced, stored or transmitted, in any form or by any means, with the prior permission in writing of the publishers, or in the case of reprographic reproduction in accordance with the terms of licences issued by the Copyright Licensing Agency. Enquiries concerning reproduction outside those terms should be sent to the publishers.

Astronomers' Observing Guides Series ISSN 1611-7360
ISBN 1-85233-751-6
Springer Science+Business Media
springeronline.com

© Springer-Verlag London Limited 2005
Printed in the Singapore

The use of registered names, trademarks, etc. in this publication does not imply, even in the absence of a specific statement, that such names are exempt from the relevant laws and regulations and therefore free for general use.

The publisher makes no representation, express or implied, with regard to the accuracy of the information contained in this book and cannot accept any legal responsibility or liability for any errors or omissions that may be made. Observing the Sun, along with a few other aspects of astronomy, can be dangerous. Neither the publisher nor the author accept any legal responsibility or liability for personal loss or injury caused, or alleged to have been caused, by any information or recommendation contained in this book.

Typeset by EXPO Holdings, Malaysia
58/3830-543210   Printed on acid-free paper   SPIN 10930076

Preface

You are holding in your hands, dear Reader, your passport to an exciting cosmic adventure – exploring the universe of double and multiple stars! These are the sky's tinted jewels and waltzing couples, and they are waiting patiently in the darkness of night to dazzle and delight you.

This is actually two books in one. The first part surveys the current state of knowledge about double stars – how they are born, evolve and interact, their significance in the cosmic scheme of things, and the valuable insights they provide into such fundamental matters as stellar masses and the ultimate fate of stars. The more we know about these fascinating objects, the more enjoyment we will ultimately derive from actually viewing them firsthand with binoculars and telescopes from our gardens or backyards or fields. As Charles Edward Barns stated in his long out-of-print classic *1001 Celestial Wonders,*

> Let me learn all that is known of them,
> Love them for the joy of loving.
> For, as a traveler in far countries
> Brings back only what he takes,
> So shall the scope of my foreknowledge
> Measure the depth of their profit and charm to me.

The second part of this book is an observing guide that tells the reader how and what to look for, examines the use of various types of telescopes and accessories (including such modern ones as video and CCD imaging devices), and offers a selection of fascinating projects suitable for novice stargazers as well as advanced amateurs. These projects range from those intended for the reader's own pleasure and edification to those that actually have the potential of contributing to the science of double star astronomy itself.

The highlight of the second section (and, indeed, perhaps of the entire book!) are two carefully compiled observing rosters – one showcasing 100 of the sky's finest double and multiple stars for viewing with telescopes from 2 to 14 inches in aperture, and the other a more extensive tabulation of 400 additional pairs for further exploration and study. Together, these two lists offer 500 selected stellar wonders for your enjoyment.

The material in this book is based on decades of studying double stars by the author as a professional observer and more than half a century as a passionate "stargazer" surveying the heavens for its visual treasures – especially striking double and multiple stars! These observations were made using everything from 7×50 and 10×50 binoculars to refracting telescopes from 2 to 30 inches (!) in aperture, reflecting telescopes from 3 to 36 inches, and catadioptric telescopes from 3.5 to 22 inches in size. It is my sincere wish that you, the reader, will find the same joy and satisfaction in becoming personally acquainted with these fascinating objects as I continue to do – even after all those years and all those telescopes!

In closing, we again quote from Barns:

> Lo, the Star-lords are assembling,
> And the banquet-board is set;
> We approach with fear and trembling.
> But we leave them with regret.

James Mullaney
Rehoboth Beach, Delaware

Acknowledgments

There are many people in the astronomical and publishing community who have helped to make this book a reality. In particular, my many friends at *Sky & Telescope* and its parent company Sky Publishing Corporation have been especially helpful in supplying information and illustrations that have appeared in the magazine. Special thanks in this regard is due to Sally MacGillivray, Rights & Permissions Manager at Sky. Dr. Nicholas Wagman, deceased director of the University of Pittsburgh's Allegheny Observatory, was both kind friend and mentor while I served on the staff there. He gave me unlimited use of its superb 13-inch Fitz-Clark refractor (and occasionally its 30-inch Brashear refractor as well!) for my early visual sky surveys and work on double stars. The late Dr. Charles Worley – long one of the world's leading double star observers with the 26-inch Clark refractor at the United States Naval Observatory and the astronomer primarily responsible for maintaining the monumental *Washington Double Star Catalog* (the *WDS*) – was also a valued source of information and inspiration to me. Since his passing, Dr. Brian Mason and his staff have taken over the task of keeping the *WDS* updated, and special thanks must be given them on behalf of the entire astronomical community for this vital service. Dr. John Watson, FRAS, and Dr. Michael Inglis, FRAS – my editors at Springer-Verlag London – have patiently guided me throughout this project and have been a sincere pleasure to work with. So too has been Louise Farkas, Senior Production Editor, at Springer New York. Finally, I wish to acknowledge my dear wife, Sharon McDonald Mullaney, who has been a constant source of help and encouragement during the long process of writing this book.

Contents

Part I All About Double and Multiple Stars

1 **Introduction** 3
   Series and Book Objectives 3
   Double and Multiple Stars Defined 3
   The Joys of "Seeing Double" 3
   Demise of a Once-Vital Field of Astronomy 6

2 **Types of Double Stars** 9
   Optical versus Physical Doubles 9
   Common Proper Motion (CPM) Pairs 11
   Visual Binaries 12
   Interferometric Binaries 14
   Astrometric Binaries 14
   Spectroscopic Binaries 16
   Eclipsing Binaries 17
   Multiple Star Systems 19
   Cataclysmic Binaries 21

3 **Astrophysics of Double Stars** 23
   Frequency and Distribution of Double Stars 23
   The Genesis of Binary Stars 24
   Determination of Stellar Masses 25
   Distances of Double Stars 25
   Double Stars and The Mass–Luminosity Relation 27
   Mass Exchange in Contact Binaries 28
   Stellar Mergers 29
   A Two-Minute Binary? 30
   Planetary Systems 31

Part II Observing Double and Multiple Stars

4 **Observing Techniques** 35
   Training the Eye 35

    Sky Conditions 37
    Star Testing and Collimation 39
    Record Keeping 41
    Personal Matters 42

5 **Tools of the Trade** 43
    Refracting Telescopes 43
    Reflecting Telescopes 44
    Catadioptric Telescopes 46
    Eyepieces 47
    Star Diagonals 48
    Barlow Lenses 49
    Micrometers 50
    Photographic, Video and CCD Imaging Systems 53
    Star Atlases and Catalogues 55
    Miscellaneous Items 56

6 **Observing Projects** 60
    Sightseeing Tour 61
    Color Studies 61
    Resolution Studies 64
    Micrometer Measurements 67
    Orbit Calculation and Plotting 71
    New Pair Survey 73
    Founding a Double Star Observers' Society 74

7 **Double and Multiple Star Observing Lists** 75
    One Hundred Showpiece Double and Multiple Stars 76

8 **Conclusion** 85
    Reporting and Sharing Observations 85
    Pleasure versus Serious Observing 87
    Aesthetic and Philosophical Considerations 88

Appendices

1 **Constellation Names and Abbreviations** 91
2 **Double Star Designations** 94
3 **Double and Multiple Star Working List** 101
4 **Telescope Limiting Magnitude and Resolution** 121
5 **The Measurement of Visual Double Stars** 123

About the Author 127

Index 129

Part I

All About Double and Multiple Stars

Chapter 1

Introduction

Series and Book Objectives

The exciting new series of observing guidebooks from Springer covers those areas of visual observational astronomy of interest to today's active amateur astronomer – from the dynamic members of our solar system to the ethereal wonders of deep space. Each volume is actually two books in one. The first provides a comprehensive overview of the class of celestial object it treats, including the latest information on both their physical properties and their significance in the grand cosmic scheme of things. The second covers observing techniques, instrumental considerations and suggested activities and programs for the telescope user.

Double and Multiple Stars Defined

We begin this volume on viewing double and multiple stars by defining just what they are: *two or more suns placed in close proximity to each other in the sky* as seen with the unaided eye, binoculars or telescopes. With the exception of stars that just happen to lie along the same line of sight but are actually far apart in space, these objects are physically (gravitationally) bound together as a system. In some cases, they are separated enough that they are simply drifting through space together, while in others they are actually orbiting around the common center of gravity of the system.

The various types of double stars are discussed in depth in the next chapter. But here an important point needs to be made. Contrary to common belief, double stars *are* deep-sky objects! Anything beyond the confines of the Sun's family is by definition in deep space. This includes single, variable and double stars, in addition to the more traditional star clusters, nebulae and galaxies. Also, throughout this book we will use the term "double star" to mean both double *and* multiple star systems.

The Joys of "Seeing Double"

Double stars are the tinted jewels and waltzing couples of the sky. Their amazing profusion, and seemingly infinite variety of color, brightness, separation and component configuration, make them fascinating as both objects of study and

Figure 1.1. Albireo (β Cygni), one of the most spectacular visual double stars in the heavens, as imaged by Austrian observer Johannes Schedler using a digital camera on an 11-inch Schmidt–Cassegrain telescope. The original 30-second color exposure shows the striking orange and blue hues of this famous pair, which is a truly magnificent sight in any telescope! Albireo can even be resolved in steadily held binoculars, which show it suspended against a superbly rich Milky Way background. Courtesy Johannes Schedler.

leisurely celestial exploration with the telescope. It is estimated that at least 80 percent of the stellar population exists as pairs and multiple groupings. Abounding among the naked-eye stars, they are visible with even the smallest of instruments on all but the worst of nights – even in bright moonlight, and through haze and heavy light pollution.

After the Moon and planets, double stars are typically the next target for the beginning stargazer before jumping into the fainter and more distant realm of clusters, nebulae and galaxies. And indeed they should be, for not only are they bright and easily found, but they are truly exciting objects! The vivid hues of bright, contrasting pairs are sights never to be forgotten. Who wouldn't remember that first view of magnificent topaz and sapphire Albireo (β Cygni) or the vivid orange and aquamarine blue of Almach (γ Andromedae) or the red and green of

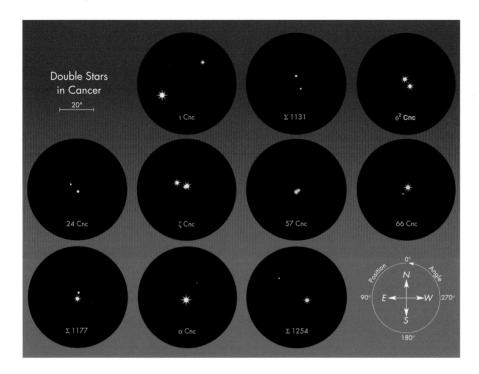

Figure 1.2. A sampling of only a few of the double stars within reach of a small telescope in the little constellation Cancer. These eyepieces sketches made by American amateur Sissy Haas, as seen through her 2.4-inch (60 mm) and 5-inch refractors, hint at the beauty and diversity displayed by visual double and multiple stars. Courtesy Sissy Haas.

Rasalgethi (α Herculis)? What about brilliant blue-white diamonds like Castor (α Geminorum) or Mizar (ζ Ursae Majoris)? And stunning multiple systems like the Double-Double (ε Lyrae) or the Trapezium (θ-1 Orionis) or Herschel's Wonder Star (β Monocerotis)?

As in other areas of amateur astronomy, part of the joy of double star observing is sharing views of these lovely "flowers in the meadows of heaven" (as Longfellow called them) with others. Their unexpected beauty typically brings gasps of astonishment and delight from first-time observers as they peer into the eyepiece. Many double star aficionados can trace their lifelong love affair with these objects to just such an encounter.

Experiences like those above often draw the observer into deeper study of these objects. Some, desiring to see as many more of them as possible, will embark on a "sightseeing tour" of the night sky. Others, wanting to permanently capture these sights, will do so by drawing what they see or imaging it photographically or electronically. Given time and patience, the orbital motions of bright pairs like Porrima γ Virginis), ξ Ursae Majoris, and Castor itself become evident. The sight of two distant suns slowly dancing about each other in the depths of interstellar space is a thrill quite beyond words. This may lead the observer to begin regular measurements of binary stars with a micrometer or other such device to follow their movements – a fascinating activity that also happens to be of great value to professional astronomers.

All of these potential needs of the double star observer are addressed in the second half of this book. **Chapter 4: Observing Techniques** provides the necessary background and personal training for observing the night sky with the unaided eye, binoculars and telescopes. **Chapter 5: Tools of the Trade** discusses instrumental considerations, including telescopes and accessories. **Chapter 6: Observing Projects** offers a number of fascinating activities for the telescope observer, ranging from beginner to advanced level. **Chapter 7: Double and Multiple Star Observing Lists** provides a delightful roster of showpiece pairs for leisurely viewing (followed by a more extensive compilation for additional study in **Appendix 3**). Finally, **Chapter 8: Conclusion** gives suggestions for sharing and reporting your observations, as well as thoughts on the ultimate purpose of your nightly vigils under the stars.

Demise of a Once-Vital Field of Astronomy

The history of double star discovery and observation is fascinating and colorful, but it lies outside the scope and purpose of this book. For those who may be inter-

Figure 1.3. One of the 20th century's greatest professional double star astronomers was George van Biesbroeck, seen here standing beside the objective of Yerkes Observatory's famous 40-inch refractor. His observing career spanned more than 70 years. Courtesy Yerkes Observatory and the University of Chicago.

Figure 1.4. The great 40-inch Clark refractor of Yerkes Observatory, located at William's Bay, Wisconsin. Long used for both visual and photographic studies of double stars, it is typical of the long-focus refractors traditionally employed in such work. The tube of this mammoth "celestial cannon" spans 65 feet in length, and it remains the largest lens-type telescope in the world today. Courtesy *Sky & Telescope* and P. K. Chen.

ested in this aspect of the subject, perhaps the finest account is that given by R.G. Aitken in his 1935 classic, *The Binary Stars*. This valuable work was reprinted by Dover Publications in New York in 1964, but has since gone out of print. It is still to be found today in many libraries, used bookstores, and over the Internet. A knowledge of the historical background of double stars definitely adds to the overall charm of actually seeing these wondrous objects.

Double star observing was a major activity and high priority of both amateur and professional astronomers a century ago. All observatory telescopes were supplied with filar micrometers for measuring binary stars, for then as now they provided our only direct means of determining stellar masses. And with the exception of those few who made – or could afford to have made – their own reflectors back then, most amateurs of the time were using small refractors in the 3- to 6-inch aperture range. The great profusion of bright double stars made them ideal targets for those wishing to explore the heavens beyond the solar system with such instruments.

But professional astronomy, which had been largely focused on measuring the positions of planets and stars, began moving outward to the realm of the clusters, nebulae and galaxies that continued to be discovered in great numbers. The photographic plate eventually replaced the eye for many studies and the spectroscope ushered in the modern era of astrophysics. Visual observation of double stars, which typically require decades or more to obtain publishable results, soon gave way to these more glamourous activities. In many cases, a few nights or weeks of observation at the telescope in these exciting new fields brought not only immediate results but significant discoveries. This caused double star work to be nearly abandoned, with only a handful of professionals and a few devoted amateur astronomers actively pursuing it by the middle of the last century.

Today we are seeing a welcome rebirth of interest in double star observing by amateur astronomers worldwide – not only for their own personal enjoyment, but also to help the few remaining professionals keep track of the multitude of known moving pairs. To aid these observers, an extensive listing of double star designations is given in Appendix 2. Based mainly on the names of the various discoverers or their catalogues or observatories, these span the nearly four-century period from the very earliest telescopic finds up to the very latest discoveries and measurements being made today.

Chapter 2

Types of Double Stars

In this chapter, we examine the amazing variety of those denizens of deep space that fall under the label "double star." And this would seem to be a good place to state right up front that double and multiple stars *are* "deep-sky objects" – a title that many observers still reserve exclusively for star clusters, nebulae and galaxies. By definition, *any* object that lies beyond the confines of our solar system is in deep space. This includes not only double and multiple stars, but single stars (both constant and variable ones) as well.

Optical versus Physical Doubles

A double star consists of two (or more) suns that appear close together in the sky when viewed with the unaided eye, binoculars or, especially, telescopes. Most of these are true *physical systems* in relative proximity to each other and bound by their mutual gravitational attraction. However, a small percentage are simply chance alignments of two objects along our line of sight. These are known as *optical doubles*. In the early days of telescopic astronomy, *all* double stars were considered to be merely optical alignments.

A majority of the objects in the double and multiple star listings in Chapter 7 and Appendix 3 are not only physical systems but actual binary stars in orbital motion about each other (see below). These lists also contain a number of well-known optical pairs – some of which are difficult for the eye to accept as being mere chance alignments. One of these is the striking naked-eye and binocular orange combo α-1/2 Capricorni, which certainly look connected but actually lie nearly 600 light-years apart in space. There are also some famous double stars whose exact status as optical or physical has been hotly debated among astronomers. An example is the naked-eye pair Mizar and Alcor (ζ +80 Ursae Majoris), at the "bend" in the Big Dipper's handle. Mizar itself was the first double star to be discovered telescopically in 1650 and is a definite physical system with a companion just 14 arc seconds away. Alcor, on the other hand, lies some 12 minutes of arc (over 700 arc seconds) from the close pair.

Traditionally, Mizar and Alcor were both considered to be 88 light-years from us. More recent measurements seemed to indicate that their real distances were 92 and 59 light-years, respectively. However, the latest precision parallaxes from the *Hipparcos* astrometry satellite place them back together again at 78 light-years. In any case, Mizar and Alcor are actually a common proper motion pair (see below) and are, therefore, definitely physically related whatever their true distances may

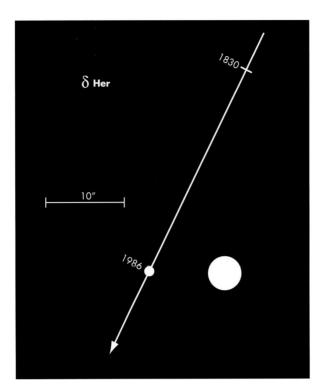

Figure 2.1. Relative motion of the apparent companion to the optical double star δ Herculis. The two objects are totally unrelated and are actually moving through space at right angles to each other. They passed at a minimum separation of 9 seconds of arc in 1960 and have been separating ever since, as shown. Such close chance alignments are relatively infrequent in the sky – most pairings are real, gravitationally bound systems.

be. Additionally, *all three stars* are spectroscopic binaries (again see below), together forming an amazing sextuple system!

An even more celebrated case involves the magnificent topaz and sapphire pair Albireo (β Cygni), considered by many observers to be the finest double star in the sky. Its accepted distance of around 400 light-years puts it just beyond the limit of reliable trigonometrically determined stellar parallaxes. Due to their angular proximity of 34 arc seconds and the radiant apparent brightness of the two stars, it was never questioned that they might not be physically related. Controversial measurements made in the mid-1980s appeared to indicate that the companion lies 1.5 times as distant as the primary. The odds that two suns so bright will be positioned so close together are something like one chance in 700 over the entire sky.

Alan MacRobert, senior editor at *Sky & Telescope*, stated in the magazine back then that "If this result holds up, Albireo is far and away the most remarkable optical double in the sky." But many, including the author, have had serious doubts about this finding. I challenge the reader to have a look for himself or herself at this famous pair, and see if you can really bring yourself to believe they are simply an accident of celestial geometry! Fortunately, the *Hipparcos* satellite has again saved the day, restoring this showpiece back to its original status as a real double star. MacRobert himself shared the good news with the author, writing, "So you can state that Albireo is a genuine binary and that it is about 380 light-years away plus or minus about 10 percent."

 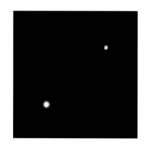

Figure 2.2. Two popular common proper motion (CPM) pairs, imaged by *Sky & Telescope* Editor-in-Chief, Rick Fienberg, using a digital camera on his 12-inch Schmidt–Cassegrain telescope. Cozy β Scorpii (left) is nicely split in a 2-inch glass, while much wider α Librae (right) can be resolved even in binoculars. Courtesy *Sky & Telescope*.

Common Proper Motion (CPM) Pairs

Many double stars in the sky (and on the lists in this book) are widely separated but drifting together through space as a couple, held in each other's embrace by their mutual gravitational attraction. These are known as *common proper motion* (or *CPM*) *pairs*. Measurements of their distances and line-of-sight (radial) velocities confirm their association. Many of these objects are undoubtedly in very slow "orbital" motion about their common center of gravity, with periods measured in thousands of years or longer.

There has been a fascinating conjecture around for quite some time now that the Sun itself may be a member of a CPM system. Its companion is pictured as being a very low luminosity red dwarf star lying at a distance of perhaps 0.25 to 0.5 light-years from us – far enough away that its gravitational effects have not yet been detected and faint enough to escape our largest telescopes. Given the current rapid advances in astronomical optics and imaging, such a discovery could well be made at any time.

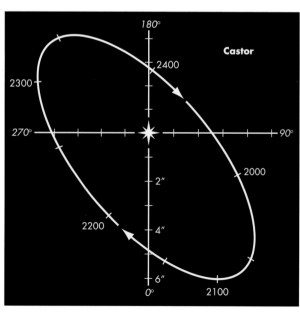

Figure 2.3. Orbit of the radiant visual binary Castor (α Geminorium). One of the first binary stars to reveal its orbital motion (to Sir William Herschel), despite its lengthy period of 470 years. A beautiful sight even in small telescopes, the two suns are currently separated by 4 seconds of arc, and are very slowly widening from their closest approach to each other of just under 2 seconds in 1963. Courtesy *Sky & Telescope*.

Figure 2.4. Digital image of the ultra-long-period visual binary Rasalgethi (α Herculis) taken by *Sky & Telescope*'s Rick Fienberg, again using his 12-inch Schmidt-Cassegrain. The original is in color and shows the pair's lovely orange and greenish-blue tints. This 5-second wide combo can be split in a 3-inch telescope on nights of steady atmospheric seeing. Courtesy *Sky & Telescope*.

Visual Binaries

A *visual binary* consists of two or more stars – visible telescopically – showing definite orbital motion as they revolve around their common center of gravity. Periods in this instance range from a few years to a few centuries. Systems with shorter periods can sometimes be detected in multiple ways (see interferometric, astrometric and spectroscopic binaries below) – the crossover point for method of detection being a function of telescope aperture/resolution and the steadiness of the atmosphere at a given time and site.

As already mentioned, observers originally thought that all double stars were mere chance alignments and paid little attention to them. It was the great English observer Sir William Herschel, accepting this to be the case, who attempted to use them to measure stellar parallaxes – the apparent displacement of a nearby star with respect to a more distant one caused by the Earth's annual motion around the Sun. While he failed in this noble attempt, his observations showed orbital motion in a number of these star pairs – most notably the bright binary Castor (or α Geminorum), proving that they are actually physically associated as gravitationally bound systems rather than mere optical associations.

As explained in more detail in the next chapter, the importance of visual binary stars is that they provide astronomers with their only direct source of stellar masses – fundamental data needed in the study of the structure and evolution of stars. Information on the instrumental requirements for observing and measuring the motion of these objects, and on computing their actual orbits, is provided in Chapters 5 and 6, respectively.

Binary star orbits are generally ellipses (except for pairs having precisely the same masses). The closest point in the companion's path around the primary is called *periastron* and the farthest point is called *apastron*. Following Kepler's well-known laws of orbital motion, the companion moves fastest at the former time and slowest at the latter time. While most pairs visible with amateur instruments take a decade or more before any orbital motion can be seen without making actual measurements, there are some striking exceptions.

One example is the close yellow and reddish double ζ Herculis. One of the fastest moving binaries within reach of amateur instruments, it has an orbital period of just 34 years and has made more than 6 revolutions since its discovery by William Herschel in 1782. Its motion reveals itself by the ease or difficulty of resolving the pair year to year as it ranges from 1.6 seconds of arc at apastron to 0.5 seconds at periastron.

Another case is the radiant pair Castor, mentioned above. It moves very slowly over much of its 470-year period, having covered just over half of its orbital circuit since its discovery in 1719 by the English astronomer J. Pound. It was an easy double for small telescopes to split until about 1960, when it began to close up noticeably as it approached periastron passage in 1963 at a separation of 1.8 seconds of arc. For a year or so before this date, a 2.4-inch (60mm) refractor at 100× showed the pair becoming noticeably tighter. Around the time of periastron, the stars came into "contact" and then their images merged – all over a period of just *several months*. As Castor opened again, this sequence was repeated in reverse order. Today, at a separation of just over 4 seconds, it is again an easy pair to see as it slowly widens to a maximum of 8 seconds in 2120.

One additional example is the famous 171-year-period binary Porrima (γ Virginis). Like Castor, it was an easy split in small telescopes until the past year or so, when it too began rapidly closing in toward its periastron passage in 2005. The pair will then be separated by just 0.4 seconds and will be unresolvable in all but the largest instruments. In small telescopes at high magnification, this object will appear as a merged elliptical blob with a rotating major axis – *the stars circling each other at an angular rate of more than 70 degrees per year*!

(Castor and Porrima appear in the showpiece listing in Chapter 7, and ζ Herculis is one of the pairs in the extended working list in Appendix 3.)

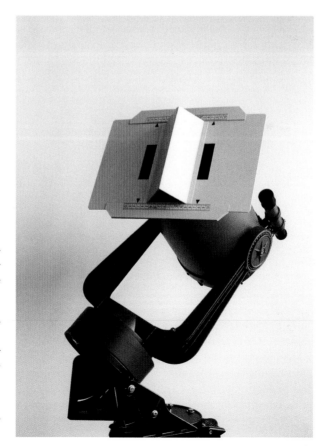

Figure 2.5. Swiss amateur astronomer Andreas Maurer has effectively doubled the resolution of his 8-inch Schmidt–Cassegrain on bright double stars using the cardboard mask and sliding screen seen here. This simple setup has converted his telescope into an optical interferometer – as it can do for any instrument having sufficient aperture. Courtesy Andreas Maurer.

Interferometric Binaries

The lower limit for resolution of visual binaries with large observatory refractors (24 to 40 inches in aperture) is somewhere between 0.25 and 0.15 seconds of arc under perfect sky conditions. Orbiting binaries certainly exist below this limit, but until relatively recently their presence could only be known in certain cases through the technique of photographic astrometry (see below).

Using the technique of interferometry has opened up an entire new class of ultra-close double stars for study, known as *interferometric binaries*. As the name implies, this procedure makes use of the interference/diffraction effects resulting from the wave nature of light. The angular resolution of a telescope (discussed in depth in Chapter 6 under the section on revising Dawes' limit) increases in direct proportion to its aperture – in other words, the bigger the instrument is, the finer detail it can see on extended objects like the Moon and planets, and the closer double stars it can split.

In theory, combining the images of an object from two different telescopes (say, for example, two 3-inch refractors) separated by a given distance – say 10 feet – results in the resolution of a scope 120 inches in diameter! In practice this direct imaging of an object is an extremely tricky operation even for skilled professionals. While still a very delicate and involved process, getting the fringe patterns of the images from the two instruments to interfere in certain ways resulting in the same resolution is easier.

The largest actual telescopic interferometer in operation today is the giant Keck binocular telescope in Hawaii, which features twin mirrors each 400 inches in diameter (twice the size of the famed 200-inch Hale reflector on Palomar Mountain in California). These behemoths are spaced 300 feet apart and are fiber-optically linked. Together, they are capable of millisecond resolution – in the range of a thousandth of a second of arc! An even more powerful instrument is the very long baseline interferometer of the Center for High Angular Resolution Astronomy in Georgia, which when fully operational will be capable of resolving binaries down to 0.0002 arc seconds apart!

A more reasonably-sized type of interferometry used by a number of professional double star observers today is the technique of *speckle interferometry*, discussed in Chapter 5. This sophisticated photographic device – typically the size of a large shoebox attached to the working end of a telescope – makes possible measurements of binaries with separations of just 0.01 arc seconds (after involved optical and mathematical treatment of the images taken). Interferometry in its various forms has opened up an exciting and dynamic realm of previously unseen fast-moving binaries for observation and study by professional astronomers.

Astrometric Binaries

Astrometry is the field of astronomy concerned with high-accuracy measurements of the positions and angular separations of stars on the sky. It is best known for the trigonometric parallaxes used to measure the distances to the stars (and ultimately other celestial objects which are calibrated upon them). This work has traditionally been done with long-focus observatory-class refractors, which provide razor-sharp images and the large, stable image scales required for measuring star positions to sub–arc second precision. However, the U.S. Naval Observatory has

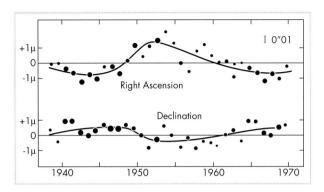

Figure 2.6. Typical plot showing perturbations (in both Right Ascension and Declination) in the proper motion of an astrometric binary having a companion with an orbital period of approximately 30 years. The error bar in the upper right hand corner is 1/100th of an arc second in size and illustrates the extreme precision required for such observations. This technique has also been successfully used in detecting planetary-mass objects orbiting nearby stars.

operated a 61-inch astrometric reflector at its field station in Ariziona for many years now, and satellites like *Hipparcos* are making astrometric measurements of unheard-of accuracy from high above the Earth's atmosphere.

Astrometric binaries are close double stars detected using the techniques of precision photographic or CCD astrometry. This involves imaging a target star against a more distant starfield, which is used as a "fixed" reference frame. This is typically done repeatedly over a period of years to allow time for the star's minute proper motion to carry it across the sky enough to be detected and measured. A relatively recent advance in this field is a device known as the "multi-channel astrometric photometer", which uses high-speed photometry and a rapidly oscillating set of slits to scan star images directly at the focal plane of the telescope. It is claimed that the device is capable of seeing the movement of the nearer stars like Sirius (α Canis Majoris) in as little as 24 hours!

When a nearby star is imaged over a period of time, both its parallax (and therefore its distance) and proper motion can be measured using any of the above techniques. A single star will move in a smooth, apparently straight line across the sky, but one with a companion (either seen or unseen) will describe a wavy path about their common center of gravity. Unseen companions detected in this way are known as astrometric binaries. These are stellar-mass objects lying below the direct visual detection threshold of a telescope, mostly because they are too close to their primary stars and/or are too faint. Many of the ultra-close pairs can also be detected by speckle and other forms of interferomery.

In addition to picking up stellar companions, some astrometric techniques are sensitive enough to detect sub-stellar-mass bodies orbiting nearby stars. In many instances, these are brown dwarfs – objects too massive to be planets but not massive enough to sustain thermonuclear reactions as real stars do. They appear to be shining by gravitational contraction. Even more exciting, in a number of cases involving nearby, high-proper-motion stars, objects of planetary mass have also apparently been detected astrometrically!

Two of the earliest and best-known of these involves Barnard's Star in Ophiuchus, and the lovely visual binary 61 Cygni (see the showpiece list in Chapter 7). The former is a red dwarf lying just 6 light-years away – the nearest star to us after the Sun and the α Centauri triple system. Its proper motion of 10 seconds of arc a year is the highest known. The latter at just 11 light-years is also among the nearest stars and was the first of the heavenly host to have its distance measured. It too has a high proper motion of 5 seconds of arc annually.

The rate of movement across the sky of these "flying stars" has made them ideal targets for astrometric study (in a few instances, even by advanced amateurs using backyard telescopes). A system of multiple planets was announced for both objects decades ago from perturbations in their motions detected on plates made with big refractors over a span of many years. In the case of Barnard's Star, these were later attributed to instrumental effects within the telescope itself, but new work suggests that there may indeed be planets here – though, curiously, *not* the same ones reported originally. Finding and verifying such bodies is apparently right at the current limit of this detection method.

The relatively long time spans required for the discovery and subsequent study of astrometric binaries has caused many astronomers to leave the field. Only a few of the world's observatories are involved in continuing such work today. Ironically, the more time that has elapsed between the first-epoch plates or images of a star and the latest ones, the more valuable the astrometric data becomes. This is due to the greater "optical leverage" – or longer motion baseline – that such data contains. Many observatories housing old plate collections are sitting on celestial gold mines of potential discoveries, just waiting for second-epoch observations to unearth them.

Spectroscopic Binaries

A *spectroscopic binary* is a star whose nature is revealed through the periodic shifting of the lines in its spectrum resulting from the well-known Doppler effect. This causes the lines to be displaced toward the blue end of the spectrum as the primary star approaches us in its orbit, and then to the red end as it recedes. If both stars are of nearly the same luminosity, a double set of lines will appear, shifting back and forth in opposite directions as the two suns do their orbital dance around their common center of gravity.

The first spectroscopic binary to be discovered (in 1889) was the primary of the well-known visual triple star system Mizar at the "bend" of the Big Dipper's

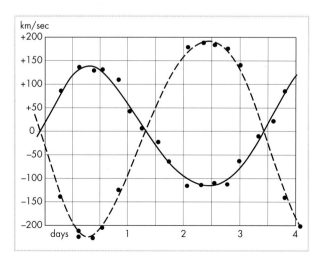

Figure 2.7. Radial velocity plot of a typical double-line spectroscopic binary – in this case, one having an orbital period of just four days. Plus values indicate motion in the line of sight away from us while minus ones reveal motion towards us, as each star revolves around the common center of gravity of the system.

handle. As previously mentioned, *all three* of its suns are spectroscopic binaries, making it an amazing sextuple system! Many of the brighter stars also have spectroscopic companions, including a majority of the double and multiple stars listed in Chapter 7 and Appendix 3. Among them is Castor in Gemini, which is a visual triple and six-star system like Mizar.

Spectroscopic binaries cover a range of component separation from that of interferometric pairs all the way down to contact binaries (see below). Some of the wider, brighter ones bridge the gap between the two methods of detection. One of the brightest spectroscopic binaries in the sky is the radiant golden sun Capella (α Aurigae). It is composed of dual *G*-type suns in a 104-day orbit, discovered spectroscopically over a century ago. The pair was later resolved using an interferometer on the 100-inch reflector at the Mount Wilson Observatory in California, and more recently directly imaged using the technique of speckle interferometry on a number of smaller instruments.

The more massive, nearby spectroscopic binaries also reveal their presence through perturbations in their proper motions and can be detected astrometrically as well as spectroscopically. If the plane of their orbit is aligned nearly edge-on to our line of sight, the two stars will periodically eclipse each other and this can be detected photometrically (see below). Thus, the three basic techniques of observational astronomy – astrometry, spectroscopy and photometry – all come together in the study of these dynamic systems. Of the three, the author has conducted both astrometric and spectroscopic research on double stars. There's no question that the latter is the most fascinating of the two approaches to the observer. Knowing that contained in that beam of light falling on the spectroscope's slit, as seen through the telescope's guiding eyepiece, are two suns rapidly orbiting each other far out in the depths of space is quite wondrous! And the added bonus of knowing that the spectrograms taken that night will within a matter of weeks or months yield their orbital elements, spurs the observer on through the longest and coldest of nights at the telescope!

It should be mentioned before moving on that there is a relatively small class of spectroscopic doubles known as *spectrum binaries* (or sometimes as "symbiotic binaries"). In these objects, the presence of an unresolved pair is revealed by a unique spectrum consisting of the lines from two stars of different temperatures – as is often the case for spectroscopic binaries in general – but strangely showing *no* Doppler shifts. These are typically indicated in catalogues with a dual spectral type, such as A3+G7, as for normal spectroscopic pairs where the lines from both stars are seen.

Eclipsing Binaries

If the orbital plane of two suns revolving about each other is close to our line of sight as seen from Earth, they will eclipse each other once every orbit – causing observable drops in the the brightness of the system. These waltzing couples are the *eclipsing binaries*. They are also sometimes referred to as "extrinsic variables" to distinguish them from genuine (intrinsic) variables that physically change in brightness due to processes within the star itself rather than as a result of celestial geometry. Nearly a fifth of the vast host of known variable stars are in reality eclipsing binaries.

The first of the eclipsing binaries to be noted by early stargazers, and by far the most famous of its class, is Algol (β Persei). Referred to since ancient times

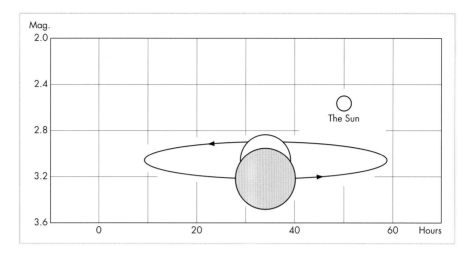

Figure 2.8. Light curve for a typical eclipsing binary system, this one having an orbital period of about three days. Here, the actual eclipse phase itself lasts just 10 hours from start to finish. Many "eclipsers" involve giant stars, and our Sun is shown for scale. The brighter eclipsing binaries make it possible to witness two suns orbiting each other far out in the depths of space with no optical aid other than the naked-eye itself.

as the "demon star" because it appears to fade and brighten (or "wink") noticeably to the unaided eye, it ranges from a normal brightness of magnitude 2.1 (outside of eclipse) to a minimum of 3.4 (during mid-eclipse) every 2.9 days. There are actually two minima. The primary drop, obvious to the naked eye, occurs when a large, dim star in the system passes in front of and partially eclipses a smaller but brighter one; there is also a much less noticeable secondary eclipse half an orbit later as the bright star passes in front of the dim one, reducing its light slightly.

The correct explanation for the variability of Algol is attributed to the young Englishman John Goodricke, who in 1782 determined the period and suggested that the regular dimming was caused by a dark star eclipsing a bright one as it revolved around it. It is truly amazing to think that when we look at Algol and other bright eclipsers that we are witnessing the mutual revolution of two suns about each other from across interstellar space using no instrument other than the unaided eye.

There are three primary types of eclipsing binaries, each named after its prototype star, and classified according to the shapes of their light curves. In "Algol-type", detached, systems the two suns are relatively wide apart and do not distort each other's shape. They are essentially spherical, having round disks with well-defined edges. Beginning and ending times of eclipses can be precisely determined from their light curve, as the edge of one star covers or uncovers the edge of the other one. Between eclipses, there is essentially no variation in brightness. They carry the standard variable star designation "EA" (for Eclipsing Algol type) and have periods ranging from a fraction of a day to many years. A well-know example of the latter is the huge supergiant system ε Aurigae, which has a period of 27 years and a primary eclipse that takes more than two years to complete.

"β Lyrae-type" systems are either detached with ellipsoidal components, or semi-detached ones in which the two suns are nearly in contact and close enough together that they *do* distort each other's shapes into ellipsoids. In the latter case, the star of greater surface brightness fills what is known as its "Lagrangian lobe" or surface – that volume around a star beyond which it cannot expand further in size without losing mass to its companion. (This critical volume is also often referred to as the "Roche lobe" or surface, derived from the famous "Roche's Limit" involving the gravitational effects of the planets of our solar system on their satellites.) Here, much of the variation seen is the result of the changing surface areas of the stars presented to us as the system as a whole rotates. The resulting light curve varies continuously, and the actual times of beginning and ending of the eclipse cannot be determined directly from it. They carry the standard variable star designation "EB" (for Eclipsing β Lyrae type), and typically have periods of a day or more.

A third class of eclipsers is the "W Ursae Majoris-type", or true contact systems, in which the two suns are roughly similar in mass and both fill their Lagrangian lobe. Their light curves resemble those of the β Lyrae stars, but the primary and secondary minima are of equal depth, and their periods are usually shorter – generally a day or less. These suns are so close to each other that their outer atmospheres are touching and in many instances exchanging mass! (See Chapter 3 for more about contact binaries.) They carry the standard variable star designation "EW" (for Eclipsing W Ursae Majoris type).

Through modern photometric studies of eclipsing binaries, an amazing amount of information can be learned about the properties of the individual stars involved (if the information is combined with spectroscopic observations.) This includes information about their sizes, shapes, masses, temperatures, limb darkening, "star spots" if present, the orbital period/eccentricity/inclination, distances apart and the rotational speed/period of each star on its axis. While largely the domain of professional astronomers, there's a very active international amateur–professional collaboration focused on the photoelectric study of such stars. For many years astronomers have visually timed the minima of these "winking stars" as part of their other variable star observations. Because this volume is devoted to visual double stars, here we will simply mention that a vast amount of information on this subject is available for those interested in such work – in the astronomical literature, over the Internet and through the numerous variable star organizations throughout the world.

Multiple Star Systems

As noted above, the majority of visual double stars have unseen astrometric, spectroscopic and/or photometric companions – making them by definition multiple stars. But as can be seen by perusing the lists in Chapter 7 and Appendix 3, many have visual companions as well. Visual triples are quite common, one of the most beautiful examples being βMonocerotis, better known as "Herschel's Wonder Star." A fair number of quadruple systems also grace the skies: two of the finest examples are ε-1/2 Lyrae – the celebrated "Double-Double" – and θ-1 Orionis, the well-known "Trapezium" in the heart of the Orion Nebula.

There are even visual multiple systems containing five, six or more members. Indeed, some objects that appear triple or quadruple in a small glass turn out to have other members that are revealed in larger apertures. In addition to the four

Figure 2.9. Photograph of the heart of the magnificent Orion Nebula (M42/M43) and its embedded multiple star, the Trapezium (θ-1 Orionis), as photographed with Lick Observatory's 120-inch reflector. In addition to the four bright stars themselves, there are many fainter members of this group (which is actually a star cluster in formation!). A 2-inch glass shows the Trapezium itself, and a 6-inch shows six stars, all set against the glowing nebulosity like diamonds on green velvet. Photograph copyright UC Regents/Lick Observatory.

bright stars of the Trapezium seen in 2- to 4-inch telescopes, there are two others that can be glimpsed in a 6-inch telescope – making this a visual sextuple system. Telescopes 12 inches and bigger show even fainter companions here, and it turns out that the Trapezium is actually a small star cluster in formation, condensing out of the surrounding nebula itself. Another example is the multiple star h3780 in Lepus. It has at least five members that can be seen in small backyard telescopes and several more in larger ones; this object has a "dual personality" as the small open star cluster NGC 2017.

The orbital motions in multiple star systems are both complex and fascinating. In the case of a quadruple like the Double-Double where the two close pairs (both under 3 seconds of arc apart) are widely separated from each other (over 200 seconds apart), there is no orbital interaction between the individual pairs themselves. (Here, the periods of the close pairs are measured in the hundreds of years, while their slow motion about each is estimated to be on the order of 10,000 centuries.) The same holds true of triple systems where there is a close pair of suns orbited by a much more distant one.

There are multiple stars, however, in which the individual components are so close to each other that it's thought that not only do their orbits interact, but also in some cases the stars actually change partners. The complex orbital mechanics and dynamics of such systems defy the imagination, and present an enormous challenge for the professional double star theorist. Actually observing such interactions is reserved for large observatory-class instruments and advanced imaging techniques. We backyard stargazers are left to simply imagine the scene within such bizarre systems – and the impact on any planets that may be present orbiting these wandering suns (see Chapter 3).

Cataclysmic Binaries

There is an entire class of contact binaries in which the sporadic ejection of mass from one star's extended atmosphere onto the other's surface sets off dramatic and sometimes violent outbursts of radiation from the receiving sun. These are the *cataclysmic binaries* (or *variables* as they are often called, due to the sudden increase in their apparent brightness over a period of days or even hours). Such binaries typically consist of a white dwarf surrounded by an accretion disk of matter thrown off by a large cool companion.

The light outbursts of these stars may amount to several magnitudes in the case of the dwarf novae, or as much as 12 magnitudes for novae themselves; both types are now known to be members of such close binary systems. Even some supernovae (those designated Type I) are believed to be the result of an exploding white dwarf in such a tightly bound system. There are also pairs in which the compact object is a neutron star or even a black hole rather than a classical white dwarf. Their presence is revealed primarily by observing their intense X-ray emissions, creating a subclass of the cataclysmics known as *X-ray binaries*. If the axis of rotation of the neutron star happens to be directed towards us, rapid pulses of radio (and in some cases light) emission are also seen; these are the famed pulsars.

There has been recent observational evidence indicating there may even be neutron star/pulsar binaries, as well as black hole binaries – and even binary systems containing one of each! A single sun of any type is surely an amazing spectacle. But gravitationally binding two or more stars in close proximity – including such diverse and bizarre objects as these – seems to be nature's preference, as we shall learn in the next chapter.

Types of Double Stars

Chapter 3

Astrophysics of Double Stars

Frequency and Distribution of Double Stars

Not only are double stars fascinating objects to view for pleasure in themselves, but they also provide professional astronomers with a vast amount of important fundamental astronomical and astrophysical data. Fortunately, there are plenty of them available for study in telescopes large and small all across the heavens.

Double stars of all types abound everywhere in our galaxy, from our own local solar neighborhood out into the spiral arms and disk of the Milky Way. And although we can't see them directly due to distance and crowding, they must surely be plentiful within the galactic nucleus, where the bulk of some 500 billion suns reside. They show themselves in great numbers in galactic clusters, as any observer who has gazed upon these glittering stellar jewel boxes knows firsthand. Despite their vastly greater distances from us (compared with those of the open clusters), double stars have been found in some of the nearer globular clusters as well. (One of these stellar beehives provides the setting for one of the most famous science fiction short stories ever written. In his classic *Nightfall*, Isaac Asimov has a civilization living on a planet orbiting a multiple star system, in which one or more suns are always in the sky and there is no night – except once in a thousand years. When nightfall finally does come, the inhabitants find the sky totally ablaze with the light of hundreds of thousands of stars!)

Double stars are also found just about everywhere on the Hertzsprung–Russell diagram, which plots the luminosity of stars against their temperature (or spectral type). It shows the position of various classes of stars along the "main sequence" and their evolutionary tracks both onto and off of it. From the radiant supergiants to the dim dwarfs, duplicity abounds. In the latter case, red dwarf and white dwarf doubles (there are even combinations consisting of both types) are very plentiful among the stellar population – most of them consisting of widely separated common proper motion pairs. There are however some striking exceptions, as we will see below.

Examples of the former that are visual doubles include the well-known red supergiants Antares (α Scorpii) and Rasalgethi (α Herculis). Another one (in this case, a spectroscopic rather than a visual binary) is Betelgeuse (α Orionis), which is so incredibly huge that it has at least one – and possibly two – companions actually orbiting *inside* of its tenuous outer atmosphere. Among the many blue supergiants that are visually double is Rigel (β Orionis). With the exception of Betelgeuse, these pairs are all attractive sights in small- to medium-sized telescopes. (See the showpiece list in Chapter 7.)

Estimates vary as to the actual percentage of the stellar population represented by double and multiple stars, but it is a large percentage indeed. Some figures run in excess of 80 percent, taking into account all of the different types of pairs discussed in the previous chapter. There are even a few astronomers who believe that *all* stars are members of multiple systems. In any case, it's certainly safe to say that at least half of them exist as pairs or larger groupings and that single stars like our Sun constitute a minority of the total. (There is also the real possibility that such stars actually have very dim and distant, as yet undiscovered, common proper motion companions, so they are really not single at all.)

The fact that so many stars occur in double and multiple systems surely says something important about how they originally formed out of the interstellar medium and also about the birth of stars in general. It is this ancient and recurring process that we examine in the next section.

The Genesis of Binary Stars

Theories of how double and multiple stars form are tied directly into theories of star birth in general. Stars form by condensing out of clouds of primarily hydrogen gas, heating up as they spin more rapidly, and becoming smaller and denser. These *protostars* glow dimly in the infrared, "shining" solely from the energy of gravitational contraction. When the temperature at their core reaches a critical value, thermonuclear reactions then commence; hydrogen is fused into helium, releasing vast stores of radiant energy into surrounding space. A star is born!

This process doesn't typically happen in isolation, but rather takes place within clusters containing dozens, hundreds, thousands and even (in the case of huge globular clusters) millions of stars. It is thought that wide double stars form by the separate collapse of two protostars that are either near each other to begin with or

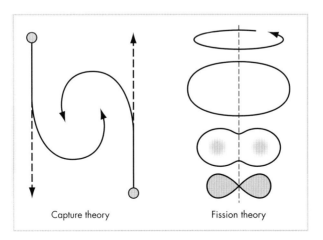

Figure 3.1. Two concepts of how double and multiple stars form. In the *capture theory* (resulting in widely separated pairs), two unrelated passing stars are caught in each other's gravitational field and end up as a loosely bound orbiting system. In the *fission theory* (resulting in close binaries), a rapidly rotating protostar cocoon splits into two nuclei that then condense into separate suns in a tightly bound orbit.

are captured into orbits around each other during a close approach in the crowded environment of their parent cluster.

It is further believed that close double stars form by a different mechanism. When a single protostar has too rapid a spin (i.e., too much angular momentum) to collapse into one sun, it splits (or fissions) instead into two separate stars caught in a tight orbital embrace. The transition between the wide pairs and the close pairs seems to occur at separations of the components ranging from 10 to 50 AUs (1 AU, or Astronomical Unit, is the average distance of the Earth from the Sun, or nearly 93,000,000 miles). For the sake of comparison, the planet Pluto's mean distance from the Sun is 39 AU. (There are also theories of orbital mechanics that seem to suggest it is possible for widely separated doubles to eventually evolve into close pairs – including the bizarre contact binaries discussed below – but these lie beyond the scope of this book.)

Determination of Stellar Masses

Binary stars offer astronomers their only *direct* means of determining a very vital and fundamental piece of astrophysical data: the masses of various types of stars. This is possible because they are in orbital motion about their common center of gravity. The principle is the same as that used when determining the masses of the Earth and Moon from the Moon's motion about the Earth, employing Sir Isaac Newton's modification of Kepler's third law of motion.

According to Newton, if any two objects are in orbital motion about each other the period of revolution increases as the distance between the objects increases, and also as the sum of the masses decreases. In the case of binary star systems, astronomers can measure both the period of revolution and the distance between the components (once the distance of the pair itself from us is known through trigonometric or other methods of parallax determination). This gives the total mass of the two objects in motion.

To find the individual masses of the stars, astronomers make use of Newton's discovery that in a system of orbiting bodies, each object actually circles around an imaginary (but physically significant) point known as the "center of gravity" of the system. By measuring the distance of each star from this center of mass, the ratio of the two masses is readily calculated. Once both the total mass and the mass ratio of a binary system are known, the individual stellar masses themselves can be determined – quite precisely for those pairs having well-defined orbits.

Distances of Double Stars

We've seen in the previous section that knowing how far double stars are from us is necessary for finding the masses of their components. Many different and ingenious methods of determining the distances to various types of celestial objects have been successfully used by astronomers. However, only one of these provides an actual direct measurement of this parameter, and upon it all the other indirect methods are ultimately calibrated – even those used to gauge the distances of galaxies. This method is known as *trigonometric parallax*.

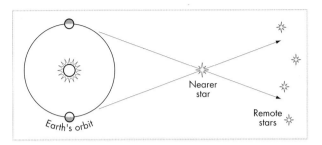

Figure 3.2. Trigonometric stellar parallax – the apparent shifting in position of a nearby star with respect to a background of more distant ones, as seen from opposite ends of the Earth's orbit. This fundamental method of directly measuring the distances to the stars by triangulation provides the calibration for all other methods of determining how far celestial objects are from us (including star clusters, nebulae and galaxies).

Parallax is based on the everyday experience here on Earth that an object seen from two different directions appears to shift in relation to the more distance background, with those closer to us shifting more than those further away. This is the principle underlying depth perception in the human eye. It was also traditionally used by surveyors in finding distances, such as that across a lake, although today this can be done with laser-ranging devices that give instantaneous values without the need to take sightings from different points.

Parallax can be applied to the Moon by making simultaneous observations of its position against the starry background from widely different locations. But when it comes to the stars, much more distant than the Moon, even a baseline as wide as the diameter of the Earth itself is insufficient to cause them to show a parallax shift. More than a century ago, astronomers hit upon the idea of using (in principle) the diameter of the Earth's orbit as a baseline for determining trigonometric stellar parallaxes. In practice, nearby stars (as revealed by their large proper motions) are photographed or otherwise imaged against a backdrop of more distance reference stars at various points around the Earth's orbit, using long-focus observatory-class refracting telescopes. Hundreds of plates are typically required to reduce observational errors and arrive at an accurate parallax for a given star.

Although the resulting shifts are measured in sub–arc seconds, they are enough to find the distances of single, double and multiple stars out to about 300 light-years or so with an accuracy of better than 10%. In the case of the very nearest stars, accuracy is within a few percentage points. Planned new orbiting astrometric satellites will extend accurate parallax measurements to several times the above range.

There is an interesting aside involving trigonometric parallaxes and double stars that should be shared here. Back in William Herschel's day, most astronomers (including Herschel) thought that double stars were mere chance alignments of two unrelated objects at different distances along the same line of sight – what are known as optical doubles. He reasoned that here was a perfect opportunity to determine stellar parallaxes by measuring the shifting of the supposed nearer star with respect to the more distant one as the Earth orbited the Sun. Instead of a parallactic shift, he found instead orbital motion – proving serendipitously that double stars are indeed real, physically related suns bound together by the force of their mutual gravity.

There are at least two additional types of "parallax" determinations (both indirect) that have been applied to finding the distances to double and multiple stars. The first of these is referred to as *dynamical parallax* and works exclusively for binary systems – in particular, those pairs with well-defined orbits. The assumption is made that the combined mass of the system is twice that of the Sun (a reasonable premise, since the combined mass of most visual binaries is typically not far from this value). Again employing Newton's generalization of Kepler's third law relating the period of revolution to the actual distance between the stars, the latter is compared to the apparent angular dimensions of the orbit to obtain a rough distance for the pair.

This preliminary distance is then combined with the apparent magnitudes of the components to find their approximate intrinsic brightness. Finally, the well-known mass–luminosity relationship (see below) is used to obtain refined values for the individual masses. In practice, this process is repeated several times until the difference in assumed and calculated masses reaches a minimum, giving a refined value of the distance itself. Dynamical parallaxes have now been determined for thousands of visual binary systems, and for those having well-defined orbits these have an accuracy of 5% or better!

A second indirect method that is widely applied to stars in general, including doubles, is that of *spectroscopic parallax*. This powerful technique uses their spectra to determine their absolute brightness or luminosity. Differences in the relative intensities of selected dark absorption lines in stars of varying luminosities (giants compared to main sequence stars like our Sun, for example) provide a sensitive indicator of intrinsic luminosity or absolute magnitude. Comparing this with the observed apparent magnitude readily gives the distance of the object. This comes from applying the well-known inverse-square law, which in essence makes it possible to determine how far away something of a given real brightness (in terms of the Sun's luminosity, which is taken as the standard) has to be from us in order to appear as faint as it does in the night sky.

The beauty of the spectroscopic parallax method is that it works irrespective of how far away an object is, in contrast to trigonometric parallax, which becomes less exact with increasing distance. If proper allowance is made for intervening interstellar dust (which makes stars appear fainter than they really are), the accuracy of this method is in the range of 10 to 15%. Once out beyond the range of precision trigonometric parallax determinations, spectroscopic parallaxes provide a far more reliable indication of a star's actual spatial distance from us than the direct method can.

Double Stars and the Mass–Luminosity Relation

It was the study of binary stars that led to the discovery by Sir Arthur Eddington of one of the most important relationships in all of astronomy. This is the well-known *Mass–Luminosity Relation*, which offers a simple but immensely valuable link between the mass of stars in general and their luminosities. In essence it states that the more massive a star is, the more intrinsically luminous it will be. And it was double stars that provided the directly measured stellar masses needed to establish this relationship.

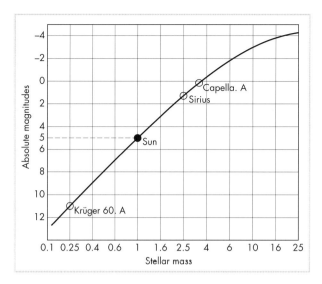

Figure 3.3. The Mass–Luminosity Relation, as originally determined theoretically by Sir Arthur Eddington. It shows that as a rule the more luminous a star is, the more massive it will be. If the absolute magnitude or luminosity of a star can be found (from knowing its distance and apparent magnitude), its mass can then be determined from this relationship. The basic unit of stellar mass is that of the Sun itself (taken to equal 1).

In practice, a star's so-called bolometric magnitude – rather than its absolute visual magnitude – is plotted against mass (again, in terms of the Sun's mass). "Bolometric" refers to the total output of a star at all wavelengths, not just in the visual spectrum. Once this has been determined by observation, a star's mass can be directly read from the plot. While the majority of stars conform to this relationship, there are exceptions. Giants and main sequence stars are in close agreement, but there are a number of supergiants that are not. White dwarfs as a class also don't agree, typically being intrinsically fainter by at least several magnitudes compared with main sequence stars of equal mass.

The Mass–Luminosity Relation, together with the Hertzsprung–Russell diagram relating stellar luminosity and temperature, have provided astronomers and astrophysicists with powerful and profound insights into the birth, evolution and death of stars. And it has been the patient observation of double and multiple stars – especially the visual binaries themselves – that has largely made this possible.

Mass Exchange in Contact Binaries

One of the most astrophysically fascinating (and bizarre!) aspects of binary stars is that presented by the *contact binaries*. As we learned in Chapter 2, these are two suns orbiting each other in such close proximity that their outer atmospheres are typically either in contact and exchanging mass with each other, or one star's outer atmosphere is spilling over onto the "surface" of the other star. The suns in such systems are enveloped by an imaginary teardrop-shaped region known as the "Lagrangian lobe" or surface (also sometimes called the "Roche lobe" or surface) – that volume surrounding each star beyond which it cannot expand further in size without then losing mass to (or towards) its companion.

Three different possible configurations of Lagrangian lobes exist (each of which can be visualized as having figure-eight cross-sections on the sky). In one configuration, neither star fills its lobe. A second configuration is where just one of

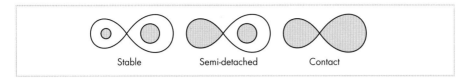

Figure 3.4. Three basic types of close binary stars. The left-hand panel shows a detached or stable system in which neither star fills its Lagrangian lobe or surface. The middle one indicates a semi-detached binary in which one of the two stars fills its lobe – typically shedding material from its outer atmosphere onto the other star, often with explosive results! The right-hand panel shows a true contact binary in which both stars completely fill their lobes and take on teardrop shapes. These dynamic binaries not only exchange mass with each other, but also eject it from the system itself in outward-flowing spirals of gas.

the stars fills its lobe. In the third type, both stars fill their lobes. Slow-moving material inside either lobe orbits around the star in that lobe, while material that moves out of either lobe escapes the gravitational pull of the star. This material may then transfer from one sun to the other as mentioned above – or, if moving fast enough, it may actually leave the system entirely.

When either star evolves towards the red giant stage in its evolution, it expands until it eventually fills its lobe completely. Its outer layers at this point assume the characteristic teardrop shape of the lobe and the Lagrangian surface now becomes the *actual* surface of the star itself! Astronomers have come to the realization that what is now the smaller star in these systems was originally the more massive one of the two. Reaching the red giant phase first, it transferred so much of its material to the other star that it then became the less massive of the two.

In fact, stellar evolution can take a normal binary through all three stages or types of Lagrangian lobes. A double star typically begins with neither component filling its lobe, and if they are widely separated from each other, they may remain this way even when the stars do eventually expand. For closer pairs, the more massive of the two suns will in time swell into a giant and completely fill its lobe. Any further expansion causes matter from its atmosphere to leave through the point of contact between the two lobes and then flow into the smaller star's lobe. While the larger star continues to lose its mass in this manner, the smaller one gains it. This, often combined with its own normal evolutionary expansion into gianthood, results in the small star eventually filling its lobe as well, resulting in a true contact binary. Both giant suns have now physically become Lagrangian surfaces; joined at their point of contact, they revolve together as a single unit like some colossal cosmic dumbbell.

Stellar Mergers

Perhaps the most sensational (and currently still controversial) aspect of contact binaries involves the possibility of the two suns in such a close-knit system merging into a single star. While actual collisions between stars are thought to be extremely rare due to the vast distances separating them (even in crowded clusters), mergers are believed to occur in the case of some ultra-tight binaries as a direct consequence of stellar evolution. (Collisions *do* happen in the case of galaxies since they are so large in proportion to the distances between them, especially in crowded clusters. But even here, while the gas and dust in these remote cities collides, the individual stars themselves pass by each other undisturbed, like phantoms in the night.)

The aspect of stellar evolution and dynamics that makes mergers appear to be possible involves mass-loss by various mechanisms from the contact binary system as a whole, bringing with it a change in the system's overall angular momentum that causes the two suns to draw closer together until merger occurs. In addition to theoretical verification that this can indeed happen, there is some observational evidence that it is actually occurring.

This comes mainly in the form of the notorious "blue stragglers" recently discovered in a number of globular clusters. These ancient swarms of stars were among the earliest parts of the Milky Way Galaxy to form and are on the order of 10 to 12 billion years old. Given their antiquity, all of the suns in these systems should have evolved into red giants by now. (Indeed, the stars in globular clusters do have an orange or ruddy hue to them when viewed in large telescopes.) But a number of intensely hot, blue stars have been found near the centers of several clusters using the *Hubble Space Telescope*. How can such young suns be present in such old, evolved systems? Some astronomers are convinced that they are the result of actual collisions of giant stars rapidly orbiting the crowded cores of globular clusters (surely a very violent event if true!). Others, however, say that the blue stragglers are direct proof that star mergers *have* occurred in aged contact binaries within these glittering stellar communities.

A Two-Minute Binary?

As we've previously seen, the orbital periods of double stars typically range from years in the case of visual and astrometric binaries to a matter of just days for eclipsing and spectroscopic binaries. In the case of contact binaries, the periods are often measured not only in days but also in *hours* – this even though one or both of the stars is often a huge giant sun. (There is also at least one obscure white dwarf pair on record that, while not a true contact binary, has been reported to have an orbital period of just 81 minutes.)

One of the shortest periods currently known for any contact binary is that of the peculiar "helium-rich" irregular variable star AM Canum Venaticorum. The original hydrogen-rich outer layers of a giant star were apparently transferred to its dwarf companion and then explosively blown away into space as the small sun went nova – a process that may have repeated itself many times. As matter continued to leave the system, both the semimajor axis and period of the orbit quickly decreased in accordance with Kepler's laws. The remaining exposed, helium-rich cores of both stars (now quite diminished in size) are revolving about each other at a dizzying rate, resulting in an orbital period of just over *17 minutes*!

But even this may not be the shortest period possible. As astronomer William Hartmann has pointed out, "Theoretically, mass loss could produce contact binaries with periods as short as 2 minutes! Somewhere in our galaxy there might be a planet in whose sky is a giant glowing figure-eight doing cartwheels like some bizarre advertising gimmick."

Planetary Systems

Just as some astronomers believe that all stars are members of double or multiple systems, others believe that all stars have planetary systems. Planets do appear, after all, to be a natural consequence of the birth process of stars. Newly forming suns spin ever more rapidly as they condense out of their nebulous cocoons and become smaller. If something doesn't slow them down, they will eventually tear themselves to pieces. As it happens, at a critical point in this process, the spin of these youthful suns suddenly drops dramatically and becomes that of a stable main sequence star. One of the theories of just how this occurs involves the ejection of rings of material from the condensing protostar, which then act as stellar "brake shoes" to slow its spin. (It is a well-known fact that if all of the planets in our own solar system were somehow pushed back into the Sun, it would then rapidly increase its spin rate and become unstable.)

The same four basic observational techniques used to observe binary stars have also been employed in the exciting search for extrasolar planets. Direct detection of planets has still not been achieved (despite several recent false alarms) due to the vast difference in brightness between them and their parent stars, but with rapid advances in imaging technology this may happen soon. The astrometric study of perturbations in the motion of stars (both their proper motion and, in the case of binaries, orbital motion) caused by the gravitational effects of planets has a long history and has turned up some promising candidates. For stars that happen to have their equatorial planes oriented towards our line of sight, precision photometry has in a number of cases revealed the definite presence of planet-sized bodies transiting a star's disk, causing very shallow eclipses of its light as they do.

Finally, the most successful technique to date has been the spectroscopic observation of variations in the radial (line of sight) velocities of stars. Detected by using some of the world's largest optical telescopes and state-of-the-art spectrographs, the minute Doppler shifts in the spectrum of a star caused by orbiting planets – pulling it first towards us and then away from us – have resulted in the discovery of well over a hundred extrasolar planets, with more found almost weekly now. Since they are heavier and their gravitational effects are easier to detect, most of these finds have been giant planets ("hot Jupiters") orbiting close-in to the star. But as techniques continue to improve, astronomers are confident that detection of Earth-sized planets is just around the corner.

So what about planets surrounding double stars? It had long been believed that due to the complex dynamics of these systems, planets could never have formed in such an environment in the first place, or if they did form they could never achieve stable orbits due to the complex gravitational interactions present. But just as all other fields of astronomy have rapidly advanced in recent years, so too has celestial mechanics (thanks largely to the advent of supercomputers and the space program). It turns out that not only can planets form in double and multiple star systems, especially in widely separated ones, but they can also achieve stable orbits. This is true even in close pairs where the planets orbit far outside the center of gravity of the system, the binary itself essentially acting as a single massive sun.

And although close-in planetary bodies may actually transfer orbits between stars – first circling one sun, then the other, and back again – a repetitive pattern brings some semblance of orbital stability.

Long before the latest theoretical modeling, there had been observational evidence of planets attending double stars. One famous case involves the lovely visual orange pair 61 Cygni, which appears in the showpiece roster in Chapter 7. Due to its large proper motion, it was suspected to lie relatively nearby – and, indeed, it became the first star to have its trigonometric parallax determined (in 1838), firmly establishing it as one of the Sun's very closest neighbors at a distance of just 11 light-years.

From astrometric studies of perturbations in both its proper motion and orbital motion, it has been known since the early 1940s that there is an unseen third body circling one of the two stars (thought to be the brighter one) with a period of roughly five years. The mass of this object turned out to be that of a big planet – just eight times that of Jupiter. More recent and extensive studies (one of which the author participated in as both telescope operator and plate measurer – the analysis being left to others!) indicate that there are *multiple perturbations* present, meaning that 61 Cygni may have an entire system of planets, as does our own Sun. (Interestingly, advanced alien astronomers observing the Sun from across space would see a "wobble" in its motion with a period of 12 years – the time it takes Jupiter, the most massive of the planets, to orbit it. More refined study would also reveal a 29-year period resulting from Saturn, the second most massive planet, and then would eventually uncover the perturbations resulting from the other, less massive planets.)

Given the amazing range and variety of sizes, types, brightnesses, component configurations and color combinations displayed by double and multiple stars (some feeling for which can be gained by looking through – and preferably at! – the objects listed and described in Chapter 7), one can only imagine the wondrous scenes visible from the surfaces of any planets orbiting them, as two or more suns fill their skies.

Part II

Observing Double and Multiple Stars

Chapter 4

Observing Techniques

It has often been said that the person behind the eyepiece is far more important than the size or type or quality of telescope being used for making celestial observations. The truth of this adage has been proven time and again; a typical example is that of a skilled observer using a much smaller instrument who sees vastly more detail on a planet like Mars than an inexperienced observer using a much larger aperture. The fact is that the eye does not work alone, but rather in conjunction with the most marvelous data processor/computer known – the human brain. It was Sir William Herschel, the greatest visual astronomer that ever lived, who said that "seeing" is an art and that as observers we must educate our eyes to really see what we are looking at in the eyepiece. This chapter is aimed at helping you get the most out of your nightly vigils under the stars, beginning first with Sir William's injunction.

Training the Eye

There are several distinct areas in which the human eye/brain combination can be trained to see better. Let's start with that of *visual acuity* – the ability to see or resolve fine detail in an image. There is no question that the more time you spend at the eyepiece, the more such detail you will eventually see. Even without any purposeful training plan in mind, the eye/brain combination will learn to search for and see ever finer detail. But this process can be considerably speeded up by a simple exercise repeated daily for a period of at least several weeks. On a piece of white paper, draw a circle, say, 3 inches in diameter. Then, using a soft pencil, randomly draw various markings within the circle, ranging from broad patchy shadings to fine lines and points. Now place the paper at the opposite side of a room at a distance of at least 20 feet or so, and begin drawing what you see using the unaided eye. Initially, only the larger markings will be visible to you. As you repeat this process over a period of time, you will be able to see more and more of them.

Taking this a step further, cut out that white disk and attach it onto a black background. Next, darken the room and illuminate the image with a low-intensity flashlight mounted in front of it. Doing this more closely simulates the view of a planetary disk seen against the night sky through the telescope. Tests have shown these exercises to improve overall visual acuity by a factor of 10! Not only will you see more detail on the Sun, Moon and planets as a result, but you will also be able to resolve much closer double stars than you were previously able to.

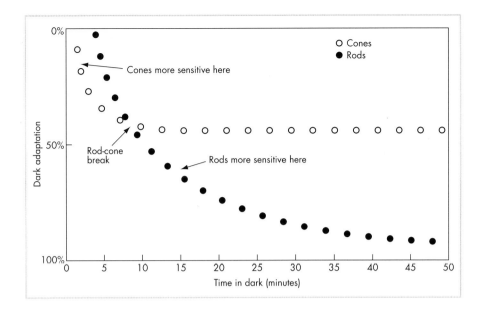

Figure 4.1. Dark adaptation times for both the color-sensitive cones (open circles) at the center of the eye and the light-sensitive rods (black dots) around the outer part of the eye. At the "rod–cone break" some 10 minutes into being in the dark, the sensitivity of the cones levels off and remains unchanged. The rods, however, continue to increase their sensitivity to low light levels; complete dark adaptation takes at least four hours! For practical purposes, the eye is essentially dark-adapted in about 30 to 40 minutes.

A second area of training the eye/brain combination involves the technique of using *averted* (or *side*) *vision* in viewing faint celestial objects. This makes use of the fact that the outer portion of the retina of the eye, which contains the receptors called *rods*, is much more sensitive to low levels of illumination than the center of the eye, which contains the receptors called *cones*. (See the discussion below involving color perception.) This explains the common experience of driving at night and seeing objects out of the corner of the eye that appear brighter than they actually are if you turn and look directly at them.

Averted vision is used in detecting faint companions to double stars and faint stars in open and globular clusters. But it is especially useful in viewing targets with low surface brightness like nebulae and galaxies, where increases in apparent brightness of 2 to 2.5 times (or an entire magnitude) have been reported. First center an object in the field of view and then look to one side of it (above or below also works); you'll see it magically increase in visibility. (Be aware there is a small dark void or "dead spot" in the fovea between the eye and ear that you may encounter in going that direction.)

One of the most dramatic examples of the affect of averted vision involves the "Blinking Planetary", a name coined by the author many years ago in a *Sky & Telescope* article about it. Also known as NGC 6826, it is located in the constellation Cygnus and is easily visible in a 3- or 4-inch glass. Here we find an obvious bluish-green 10th-magnitude nebulosity some 27 seconds of arc in size surrounding a 9th-magnitude star. Staring directly at the star, there is no sign of the nebulosity itself. On switching to averted vision, the nebulosity instantly appears and is so bright that it drowns out the central star. Alternating back and forth between direct and averted vision results in an amazing apparent blinking effect.

A third important area involving the eye/brain combination is that of *color perception*. At first glance, the stars all appear to be white. But upon closer inspection, differences in tint among the brighter stars reveal themselves. The lovely contrasting hues of ruddy Betelgeuse and blue-white Rigel in the constellation Orion is one striking example in the winter sky. Another can be found in the spring and summer sky by comparing blue-white Vega in Lyra and orange Arcturus in Bootes. Indeed, the sky is alive with color once the eye has been trained to see it. Star color, by the way, is primarily an indication of surface temperature: ruddy ones are relatively cool, while bluish ones are quite hot. Yellow and orange suns fall between these extremes.

While the rods in the edge of the eye are light sensitive, they are essentially colorblind. Thus, for viewing the tints of stars or other celestial wonders, direct vision is employed, making use of the color-sensitive cones at the center of the eye. Stare directly at an object to perceive its color and off to the side to see it appear brighter (unless it is already a bright target like a planet or 1st-magnitude star). It should be mentioned here that there is a peculiar phenomenon known as the "Purkinje effect" that results from staring at red stars: they appear to increase in brightness as you watch them.

One final note concerning preparation of the eye to see is that of *dark adaptation*. It is obvious that the eyes needs time to adjust to the dark after coming out of a brightly lit room. Two factors are at play here. One is the dilation of the pupils themselves, which begins immediately upon entering the dark and continues for several minutes. The other involves the actual chemistry of the eye, as the hormone rhodopsin (often called "visual purple") stimulates the sensitivity of the rods to low levels of illumination. The combined result is that night vision continues to improve noticeably for perhaps half an hour or so. This is why the sky looks black on first going outside, but later looks gray as you fully adjust to the dark. In the first instance, it is a contrast effect, and in the second instance, the eye has become sensitive to stray light, light pollution and the natural airglow of the sky itself that were not seen initially.

Stargazers typically begin their observing sessions by viewing bright objects like the Moon and planets and then moving to fainter ones, giving the eye time to gradually dark-adapt naturally. This is mainly of value in observing the dimmer deep-sky objects like nebulae and galaxies. Double stars themselves are generally so bright that they can be seen to advantage virtually immediately upon going to the telescope. Exceptions are faint pairs and dim companions to brighter stars (where the radiance of the primary often destroys the effect of dark adaptation). White light causes the eye to lose its dark adaptation but red light preserves it, making it standard procedure to use red illumination to read star charts and make notes at the eyepiece.

Sky Conditions

A number of atmospheric and related factors affect the visibility of celestial objects at the telescope. In the case of double and multiple stars, the most important of these is atmospheric turbulence, or *seeing*, which is an indication of the steadiness of the image. On some nights, the air is so unsteady (or "boiling") that star images

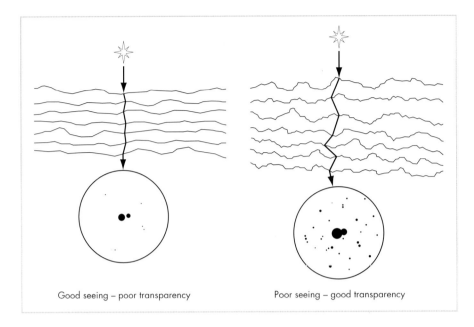

Figure 4.2. On nights of good "seeing", or atmospheric steadiness, the air above the observer is tranquil and lies in relatively smooth layers, often resulting in somewhat hazy skies. This allows starlight to pass through little disturbed and produces sharp images at the eyepiece. In poor seeing, the atmosphere is very turbulent, typically bringing with it crystal clear skies (and an ideal time to view faint objects). But the resulting star images are often blurred, shimmering balls of light, making such nights undesirable for splitting close double stars (as well as for seeing fine detail on the Moon and planets).

appear as big puffy, shimmering balls and detail on the Moon and planets is all but nonexistent. This typically happens on nights of high *transparency* – those nights having crystal clear skies, with the air overhead in a state of rapid motion and agitation. On other nights, star images are nearly pinpoints with virtually no motion, and fine detail stands out on the Moon and planets like an artist's etching. Such nights are often hazy and/or muggy, indicating stagnant tranquil air over the observer's head.

One of the most dramatic and revealing accounts of the changing effects of seeing conditions upon celestial objects comes from the great double star observer, S.W. Burnham, in the following account of the pair Sirius (α Canis Majoris): "An object glass of 6 inches one night will show the companion to Sirius perfectly: on the next night, just as good in every respect, so far as one can tell with the unaided eye, the largest telescope in the world will show no more trace of the small star than if it had been blotted out of existence."

Various "seeing scales" have long been employed by observers to quantify the state of atmospheric steadiness. One of the most common of these uses a 1 to 5 numerical scale, where 1 indicates hopelessly turbulent blurred images and 5 indicates stationary razor-sharp ones. The number 3 denotes average conditions. Others prefer a 1 to 10 system, where 1 again represents very poor seeing and 10 virtually perfect seeing. In some schemes, the numerical sequence is reversed, with lower numbers indicating better and higher numbers poorer seeing. Casual double star observing can be done in all but the worst of seeing conditions, but projects like the of micrometer measurement of binary systems (described in Chapter 6) require the very best seeing possible.

Among other factors affecting telescopic image quality is that known as "local seeing" – thermal conditions in and around the telescope itself. Heat radiating from driveways, walks and streets, houses and other structures (especially on nights following hot days) plays a significant role. This is why observing from grassy areas away from buildings gives the best results.

The cooling of the telescope optics and tube assembly is especially critical to achieving sharp images. Depending on the season of the year, it may take up to an hour or more for the optics (especially the primary mirror in larger reflectors) to reach equilibrium with the cooling night air. During this cool-down process, air currents within the telescope tube itself can play absolute havoc with image quality, no matter how good the atmospheric seeing is – even in closed systems like the popular Schmidt-Cassegrain. Surprisingly, even the heat radiating from the observer's body can be a concern here, particularly with telescopes having open-tubed truss designs.

Star Testing and Collimation

For casual observation of double stars, even a telescope of mediocre optical quality can provide acceptable views, but for more demanding work, such as resolving close pairs or making micrometer measurements, high optical quality is a must. The condition of a telescope's optics and their all-important optical alignment can readily be determined by a simple test using a star itself. Known as the *extrafocal image test*, this involves looking at the image of a star, both inside and outside of focus using a medium- to high-power eyepiece. An ideal star for this purpose is 2nd-magnitude Polaris (α Ursae Minoris), which is neither too bright nor too faint, and has the added advantage of not moving at the eyepiece due to the rotation of the Earth.

A telescope having first-class (or "diffraction-limited") optics in perfect alignment (or *collimation*) will show identical circular disks of light with a pattern of faint concentric dark interference rings on either side of focus, as the eyepiece is

  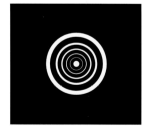

Figure 4.3. The out-of-focus (or extrafocal) image of a star can reveal many things about a telescope's optics (as well as its thermal environment and the state of the atmosphere) – in this case, alignment of the optical components. The image in the left-hand panel reveals gross misalignment. That in the middle one shows moderate misalignment (still enough to degrade image quality), while the right-hand panel indicates perfectly aligned optics.

racked in and out. These rings should be uniformly spaced and of even intensity; if not, this indicates uneven zones in the optical figure, a condition known as "spherical aberration". A "shaggy" look to the ring indicates a rough finish to the glass rather than the desired smooth one. If the extrafocal images are triangular rather than circular, this indicates that the lens or mirror is pinched in its cell. Elliptical images that rotate 90 degrees on either side of focus are the most to be feared since they reveal the serious optical defect known as "astigmatism", or a warping of the glass itself. However, both astigmatism in the observer's own eye and optical misalignment can also produce the same effect. If you wear eyeglasses to correct for astigmatism, you should definitely leave them on while conducting such tests (as opposed to the normal procedure of removing your glasses for comfort during observing and correcting by focusing the telescope). There have been cases where a bad eyepiece actually caused an astigmatic image rather than the primary optical element itself. This can easily be diagnosed by simply turning the eyepiece while looking at the extrafocal star image; if the major axis of the ellipse turns with it, the problem resides in the eyepiece. (Similarly, if turning your head while looking into the eyepiece causes the ellipse to rotate, the astigmatism is in your eye and not the telescope.)

Optical collimation is an entire subject unto itself. The term describes that condition where all the optical elements in a telescope system are aligned onto the same optical axis, to very high angular tolerances. The actual process of collimation depends on the type of instrument. Refractors are typically permanently collimated and in most cases no provision is made for adjusting it. Likewise, Maksutov systems are also usually permanently collimated. Reflectors are normally aligned using the three push-pull adjusting screws on the back of the primary mirror cell, but they occasionally also require adjustment of the secondary mirror with its corresponding screws. Schmidt-Cassegrain systems can only be aligned by adjusting the secondary mirror attached to the corrector plate using three push-pull screws. Instructions for collimation are normally included in the instruction manuals provided with all commercially made telescopes today, and many companies also supply a simple "collimation eyepiece" to aid in the process.

While rough optical collimation can be achieved by looking though the eyepiece holder and making adjustments until the images of the different elements appear aligned and concentric with each other, precise collimation is best done on an actual star image. (Some laser-alignment devices have appeared on the market in recent years that make possible precision collimation without actually looking at a star image.) When the extrafocal image shows perfectly concentric rings, the system is collimated. If they are bunched together and skewed to one side of the image, then adjustment is needed. With practice and some patience, the trial-and-error process of bringing the rings into concentricity can be accomplished in a matter of minutes.

It should be pointed out here that the extrafocal image test also makes it possible to judge something of the seeing conditions, ranging from those local ones within and immediately above the telescope to the upper atmosphere itself. Undulating waves and flashing patterns of light crossing over the diffraction-ring pattern can tell much about the steadiness of the atmosphere, as well as the telescope's thermal state and environment. Careful focusing will reveal patterns floating at different levels above the telescope's light path and (again with practice) can tell the observer if the disturbances are in the atmosphere or in the vicinity of the telescope itself.

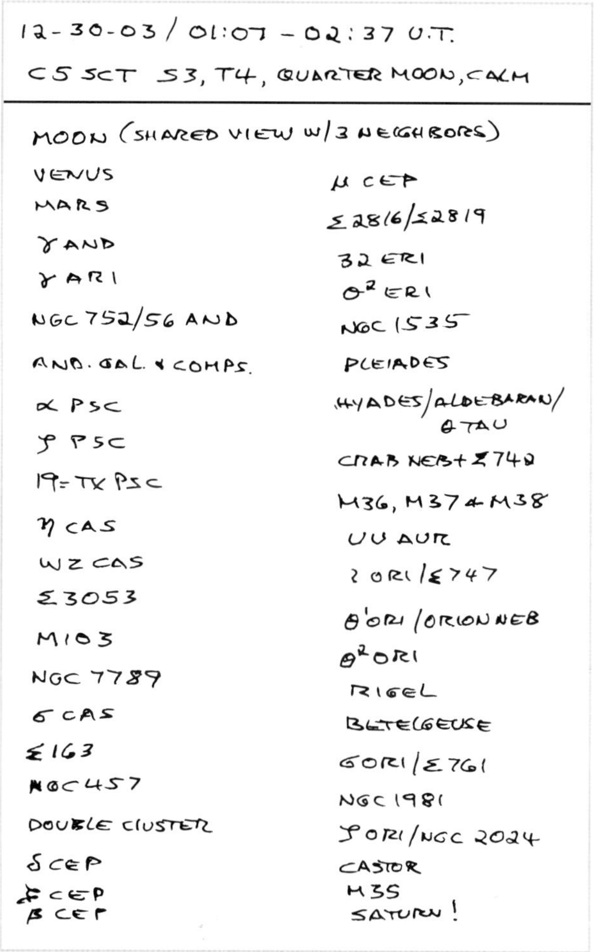

12-30-03 / 01:07 – 02:37 U.T.
C5 SCT S3, T4, QUARTER MOON, CALM

MOON (SHARED VIEW W/3 NEIGHBORS)

VENUS	μ CEP
MARS	Σ2816/Σ2819
γ AND	32 ERI
γ ARI	θ² ERI
NGC 752/56 AND	NGC 1535
AND. GAL. & COMPS.	PLEIADES
α PSC	HYADES/ALDEBARAN/θ TAU
ψ PSC	CRAB NEB + Σ742
19 = TX PSC	M36, M37 & M38
η CAS	UU AUR
WZ CAS	λ ORI / Σ747
Σ3053	θ¹ ORI / ORION NEB
M103	θ² ORI
NGC 7789	RIGEL
σ CAS	BETELGEUSE
Σ163	σ ORI / Σ761
NGC 457	NGC 1981
DOUBLE CLUSTER	ζ ORI / NGC 2024
δ CEP	CASTOR
ξ CEP	M35
β CEP	SATURN!

Figure 4.4. A busy night's entry from the author's personal observing logbook. The date and times are given in Universal Time (U.T.) – that of the Greenwich, England, time zone. A 5-inch Schmidt-Cassegrain catadioptric telescope (C5 SCT) was used under conditions of average seeing (S) and good transparency (T), and the sky was brightened by the light of a first-quarter Moon. All of the targets viewed on this particular night were (are) celestial showpieces! Normally, more time should be given to viewing fewer objects than in this case in order to fully enjoy and appreciate the cosmic pageantry.

Record Keeping

The anuals of both amateur and professional astronomy attest to the personal and scientific value of keeping records of our nightly vigils beneath the stars. From a personal perspective, an account of what has been seen each night can have a surprising impact as we look back over the years to our first views of this or that celestial wonder, or to the time we shared a first look at the Moon or Jupiter or Saturn or other sight with loved ones, friends and even total strangers. Our eyepiece impressions recorded on paper (written and/or sketched) or perhaps audio taped can provide many hours of nostalgic pleasure in years to come.

For casual double star observers, there is a definite aesthetic value in logging the various pairs seen (especially the showpiece ones), including their colors and component configurations and degree of visibility in a given telescope at a particular magnification under various sky conditions. In more serious undertakings like micrometer measuring, actual data needs to be recorded for later analysis and submission. In either instance, the information in your logbook should include the following: the date and beginning and end times of your observations (preferably Universal Time/Date); telescope size, type and brand used; magnification(s) employed; sky conditions (seeing and transparency on a 1 to 5 or 1 to 10 scale, along with notes on passing clouds, haze, moonlight and other sources of light pollution); and finally a description (accompanied by a sketch, if you're so inclined) of each object seen.

Two important points should be borne in mind regarding record keeping at the telescope. First, keep the amount of time you spend logging your observations to a minimum (and use a red light to preserve your dark adaptation when you do). Some observers spend far more time writing about what they see at the eyepiece than they spend actually seeing it! Secondly, even a negative observation (for example, in searching for a comet or looking for supernovae in other galaxies) can have value. Often has the call gone out to the astronomical community in the various magazines, journals and electronic media asking if anyone happened to be looking at a certain object or part of the sky on a particular date and at a particular time. (This frequently happens in attempting to determine when a nova first rose to visibility.) If you happened to be looking at the right place at the right time but noted nothing unusual in your observing log, that is still a fact of real importance to researchers.

Personal Matters

There are a number of other factors that impact the overall success of an observing session at the telescope. One concerns proper dress. This is of particular importance in the cold winter months of the year, when observers often experience subzero temperatures at night. It is impossible to be effective at the eyepiece, or to even just enjoy the views, when you're half frozen to death! Proper protection of the head, hands and feet are especially critical during such times, and several layers of clothing are recommended. During the summer months the opposite problem occurs, as observers attempt to stay cool. In addition to very short nights at this time of year, there's the added annoyance of flying insects and optics-fogging humidity and dew (see the discussion on dew caps and heated eyepieces in Chapter 5).

Another concern is proper posture at the telescope. It has been repeatedly shown that the eye sees more detail in a comfortably seated position than when standing, twisting or bending at the eyepiece. If you must stand, be sure that the eyepiece/focuser is at a position where you don't have to turn and strain your neck and head to look into it. And while not as critical, the same goes for positioning finder scopes where they can be reached without undue contortions.

Proper rest and diet both play a role in experiencing a pleasurable observing session. Attempting to stargaze when you're physically or mentally exhausted is guaranteed to leave you frustrated and maybe looking for a buyer for your prized telescope. Even a brief catnap before going out to observe after a hectic day is a big help here. Many heavy foods can leave you feeling sluggish and unable to function alertly at the telescope. It's much better to eat after you're done stargazing, especially since most observers find themselves famished then (particularly on cold nights!). Various liquid refreshments such as tea, coffee and hot chocolate can provide warmth and a needed energy boost. And while alcoholic drinks like wine do dilate the pupils and technically let in more light, they adversely affect the chemistry of the eye. This reduces its ability to see fine detail on the Moon and planets, resolve close double stars and see faint objects like galaxies.

Chapter 5

Tools of the Trade

In this chapter we examine the various types of telescopes and accessories employed by the modern amateur astronomer, with particular emphasis in the latter case on those devices of direct interest to double star observers. While it is true that literally hundreds of these stars can be seen around the sky with just a pair of binoculars – and a good number with the unaided eye alone – a telescope is necessary to really appreciate and enjoy this class of deep-sky object.

Refracting Telescopes

The oldest and most basic of astronomical telescopes is the lens type, or *refractor*. In its earliest form, single lenses of different sizes and focal lengths were fitted to the opposite ends of a tube through which the observer looked directly at a celestial target. But single pieces of glass act like weak prisms, and so these crude instruments suffered from an optical defect known as *chromatic aberration* and gave dismally poor images. This led to the development of the *achromatic* refractor, in which two elements of different types of glass were combined to form the primary lens, greatly reducing the false color seen with a single lens instrument.

These traditional achromatic telescopes require long focal lengths (typically, focal ratios of f/10 to f/15 or more) in order to keep color aberration within acceptable limits. This results in long and sometimes unwieldy tube lengths but has the

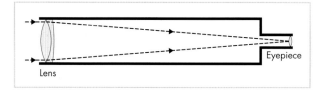

Figure 5.1. The optical configuration and light-path of the classical achromatic refracting telescope, which employs a double-lens objective. Apochromatic refractors having objectives composed of three or more elements are also widely used by today's amateur astronomers. These more highly color-corrected systems are mainly available in apertures of 6 inches and under and are much more costly than a traditional refractor.

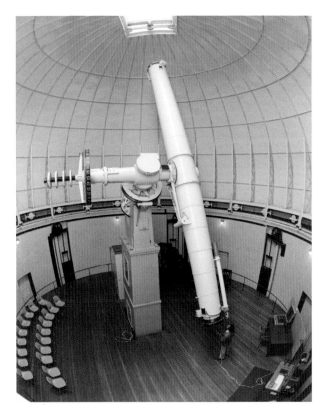

Figure 5.2. The 26-inch Clark refracting telescope of the United States Naval Observatory in Washington, DC. In active use for more than a century in the measurement of visual binary stars, this is the famous instrument which Asaph Hall used to discover the two satellites of Mars in 1877. Courtesy U.S. Naval Observatory Library.

advantages of large image scale and higher magnifications. This has made such instruments the choice of both professional and amateur astronomers for double star work, especially the micrometer measurement of close binaries. Virtually all of the world's largest refractors are basic double-lens achromats, including the great 40-inch glass at Yerkes, the 36-inch glass at Lick and the 33-inch glass at Meudon. The author has conducted visual and photographic observations with the 30-inch refractor at the Allegheny Observatory, as well as with a host of smaller refractors from 2- to 24-inches in aperture over the years and can personally attest to the superb image quality these telescopes are capable of delivering.

In recent years, the *apochromatic* refractor has taken center stage, mainly among amateur astronomers. These compact instruments have 3- (and sometimes 4-) element objectives and fast focal ratios of f/4 to f/8, providing excellent color correction in a short, highly portable tube. With their low powers and wide fields of view, these are wonderful telescopes for all types of deep-sky observing, except when high magnifications are needed, as for resolving close doubles and measuring binaries. For these purposes, a *Barlow lens* (see below) is typically employed to extend the focal length.

Reflecting Telescopes

In an effort to bypass the color aberration inherent in the simple objectives of early refractors, the *Newtonian reflector*, or mirror-type telescope, was invented. This

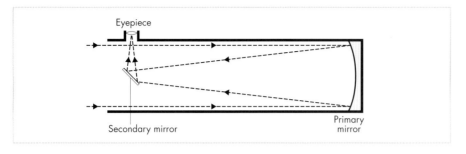

Figure 5.3. The optical configuration and light-path of the classical Newtonian reflecting telescope. A parabolic primary mirror reflects the light onto a small flat secondary one, which directs it to a focus at the side of the tube. Most of the world's large research telescopes are various forms of reflectors. This is partly because their huge mirrors can be supported from behind (instead of around the edge as with refractors) and also because the glass itself does not need to be of "optical" quality since the light merely reflects off its polished and coated surface rather than passing through the glass itself (again, as in the case of refractors).

form uses a concave parabolic primary mirror at the bottom of the telescope tube to collect incoming light and reflect it back to a focus near the top of the open tube. Before it reaches focus, the beam is intercepted by a small flat secondary mirror placed at a 45-degree angle, directing it to the side of the tube where it can be examined with an eyepiece.

While it is true that the secondary mirror does obstruct some of the incoming light, and causes diffraction effects that reduce image contrast (the actual degree depending on its size relative to the primary mirror), for all practical purposes a reflector with good optics will provide essentially diffraction-limited performance – especially long-focus ones. And its images are totally free of spurious color, since light doesn't pass through any glass (except in the eyepiece), but rather bounces off of an aluminum or other reflective coating vacuum-deposited onto the optical surfaces.

Focal ratios with reflectors typically range from f/4 (mainly in the popular Dobsonian-mounted form) in short-focus systems to f/10 in long-focus ones. (At the latter focal ratio and above, the primary mirror can actually be left spherical. Such instruments typically provide refractor-like performance due mainly to their small secondary mirrors – again, without the color concerns of a lens-type system.)

Another form of reflector is the *Cassegrain* telescope, in which a convex secondary mirror returns the converging light from the primary back through a hole in its center. The eyepiece is positioned behind the primary mirror, and the observer looks straight through the telescope as with a refractor. The secondary's convex curve amplifies the effective focal ratio of the system, with typical values ranging from f/10 to f/20 or even greater. Thus a very long focal length instrument can be fitted inside a very short tube, making these very compact instruments for their aperture. (Some reflectors actually have interchangeable secondaries, making it possible to switch between the Newtonian and Cassegrain mode, the former providing low-power/wide-field views and the latter high-power/large image-scale ones.)

Image quality in a classic Cassegrain reflector is typically not as good as in a Newtonian reflector and is definitely inferior to that in a refractor owing to a variety of aberrations, including *coma* and *field curvature*. This has led to the development of a number of exotic variations, the most effective and widely used being the *Ritchey-Chretien* design, which provides a large flat field with sharp images to the edges. Nearly all of the world's big reflectors today are "R-Cs", as they are called, including the giant 400-inch Keck twin telescopes in Hawaii. So too is the best-known telescope of them all – the Hubble Space Telescope. However, the famed 200-inch Hale reflector on Palomar Mountain in California is essentially a classical Newtonian/Cassegrain. (This telescope has actually never been used at a Newtonian focus, but instead at the "prime focus", where in times past the observer rode in a circular cage high above the primary mirror to make the observations.) Like all large research reflectors today, it is used mainly at the Cassegrain focus – as well as at the "Nasmyth" and/or "Coude" ones, where the light is intercepted at the bottom of the tube by another mirror before entering the primary's hole and directed to a number of observing positions outside of the telescope itself.

Catadioptric Telescopes

The type of telescope most widely used by amateur astronomers today is the *catadioptric*, or lens-mirror system. A relatively recent development, these instruments combine a very fast (f/1 or f/2) spherical primary mirror with either a steeply-curved meniscus lens (the Maksutov-Cassegrain) or a thin, nearly flat aspheric lens (the Schmidt–Cassegrain) to correct the coma and spherical aberration inherent in such fast optical systems. Effective focal ratios are typically f/10 to f/14, providing long focal length performance in an incredibly short tube.

These compound systems combine the best features of the refractor and the reflector, packaging them into a very compact and highly portable telescope. (The tube assembly for a 12- or 14-inch scope can actually fit on the back seat of a medium-sized car!) The 8-inch Schmidt-Cassegrain, in particular, has become the most widely used and popular telescope in the world today (with the exception of the common 60mm or "2.4-inch" imported refractor sold everywhere), according to *Sky & Telescope* magazine.

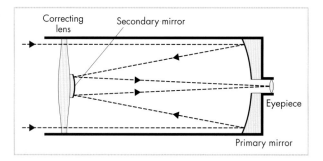

Figure 5.4. The optical configuration and light-path of a catadioptric telescope. The form seen here is the popular and widely used Schmidt–Cassegrain, which employs a thin aspheric corrector plate or "lens" to eliminate the aberrations of the spherical primary mirror. The other main type of compound telescope in use today is the Maksutov-Cassegrain, which substitutes a thick, steeply curved meniscus lens for the corrector plate.

For casual double star observing – and even for undertaking many of the observing projects discussed in the next chapter – virtually *any* size and type of telescope can be used if it is optically sound. But for the demanding and important task of measuring double stars, the refractor long reigned supreme. However, a number of observers both present and past have employed long-focus Newtonian and Cassegrain reflectors very effectively for this purpose. Among them was the noted double star astronomer, George van Biesbroeck. Although he used the Yerkes 40-inch refractor for most of his work, he also made many measurements with the 82-inch reflector at McDonald Observatory. In more recent years, a number of amateurs have made high-quality measurements of visual binaries with 8- to 12-inch reflectors using filar and other types of micrometers.

The use of catadioptric telescopes for such work is becoming more widespread, the most frequently employed instrument being the 8-inch Schmidt-Cassegrain. The advantages and disadvantages of these systems are discussed in Chapter 6 under the section on measuring double stars. There you will also read about a remarkable amateur who is making professional-quality measurements of binaries using a 3.5-inch (90mm) Maksutov-Cassegrain.

Eyepieces

It is the eyepiece in a telescope that does the actual magnifying of the image, after the objective lens or primary mirror has collected the light from a celestial object and brought it to a focus. It also happens to be the element in the optical train that is most often overlooked as the source of good or bad overall system performance. An eyepiece can make or break even the best of telescopes.

There are a plethora of eyepiece designs on the market, ranging from inexpensive, primitive two-element oculars to multiple-element units containing up to seven or eight individual lenses and costing as much as some telescopes do themselves! In general, an eyepiece should be well corrected for color and other aberrations, and have as wide and flat a field (i.e., little or no curvature) of view as possible. Two of the most common types long used by observers are the *Kellner* and the *Erfle*. Among the most widely used forms today are the *orthoscopic* and the *Plossl* designs, which are not only affordable but also give excellent performance for both casual and serious observing. Of the many wide-field and super-wide-field designs available, the *Nagler* leads the pack with its incredible "space-walk" performance.

A few words are in order about the fields of view of eyepieces. There are two basic parameters involved. One is an eyepiece's *apparent field* – the angular extent in degrees seen looking through it against a bright surface like the daytime sky. Values range from only 30 or 40 degrees to a whopping 85 degrees, depending on type and design/make. Most eyepieces in use today typically have 50-degree apparent fields. The other parameter is the *actual field* an eyepiece delivers, which is a function of both its apparent field and the magnification it produces with a given telescope (see below). It is found by simply dividing the former by the latter: actual field = apparent field/power. Thus, an eyepiece having an apparent field of 50 degrees and magnifying 50 times (or 50×) provides an actual field of one degree. When the power is doubled to 100×, the actual field drops to only half a degree, at 200× to a quarter of a degree, and so on. The higher the magnification, the smaller the amount of sky that's seen. It should be pointed out here that one degree (1°) contains 60 minutes (60') of arc and each minute contains 60 seconds (60") of arc. The Moon at its average distance is about half a degree or 30' in apparent size when seen with the unaided eye. So a telescope/eyepiece combination that delivers a 1-degree actual field can fit two full Moons into its view.

Determining the actual magnification that an eyepiece gives on a telescope is equally straightforward. The power (×) of a given telescope/eyepiece combination is found by dividing the focal length of the telescope by the focal length of the eyepiece. The focal length itself is the distance (in inches or millimeters) from a lens or mirror to its focus. For a telescope it is generally capitalized ("Focal Length"), while for an eyepiece it is in lower case ("focal length"). Some prefer to use "F.L." and "f.l." to differentiate between the two. Thus, Power = Focal Length/focal length (or X = F.L./f.l.). A telescope having a 50-inch (or 1250mm) Focal Length, used with a 1-inch (25mm) focal length eyepiece, results in a magnification of 50×; with a 0.5-inch (12.5mm) eyepiece, the power is 100×, and so on.

Incidentally, eyepieces come mounted in several different sized barrel diameters today. The *subdiameter size* 0.965" ocular is most often found on inexpensive telescopes, in particular those imported from Japan and other countries in the Far East. These oculars typically have very limited fields of view, inferior optical quality, and poor eye relief (the distance from the eyepiece's eye-lens where the entire field of view can be seen). The *American standard size* 1.25" diameter is the most widely used, its larger barrel providing room for multi-element lenses that result in excellent optical quality, roomy fields of view and comfortable eye relief. And finally there's the *giant* or *jumbo size* 2" eyepiece barrel used for some of today's most sophisticated, ultra-wide-field oculars.

Virtually all sizes, types and makes of eyepiece can and have been used for double star observing. However, the better the eyepiece, generally the more pleasing the views will be. In particular, one having good color correction should be used for studying the tints of doubles visually or in attempting to photograph them. For measuring pairs, an eyepiece providing high image contrast, minimal field curvature (mainly for wider doubles) and lack of internal reflections is desired. Concerning this last feature, all eyepieces – and, indeed, all refractive glass surfaces, especially objective lenses – should have antireflective coatings. (Some inexpensive eyepieces produce so many internal reflections or "ghosts" that they are considered to be "haunted"!)

One of the best descriptions available of the various eyepiece designs and types, and their relative performances, can be found in Philip Harrington's monumental work, *Star Ware*, published by John Wiley & Sons. Currently in its third edition and widely available, it also contains probably the best information in print on all aspects of telescopes and their accessories, with particular emphasis on the multitude of commercially made instruments flooding today's market. His in-depth coverage of these topics goes far beyond the intended scope of this book. For anyone planning to purchase a telescope, a set of eyepieces or any astronomical equipment for that matter, Harrington's opus should be considered "required" reading before doing so.

Star Diagonals

Many pictures show an observer peering straight through a long refracting telescope. This image is quite misleading. For objects on the ground or low in the sky, this may work satisfactorily. However, most astronomical targets are positioned at high altitude angles in the sky, all the way up to the overhead point itself, and it is virtually impossible to bend the neck to view them straight through. (This isn't a concern using a Newtonian reflector, because observing is conducted from the side and near the top of the telescope tube.) This problem is also common to Cassegrain and catadioptric telescopes, where observing is done at the back end of the instrument as with a refractor. To overcome it, the *star diagonal* was invented.

This device basically consists of two tubes joined at right angles to each other in a housing containing either a precision right-angle prism or first-surface mirror, one tube fitting into the focuser and the other accepting the eyepiece. The converging light beam from the primary mirror or objective is turned 90 degrees to the optical axis by the star diagonal, where the image can then be examined in comfort without contorting the neck.

It should be noted that a diagonal produces a mirror-image of what is being viewed, so that objects appear right side up but reversed left to right. This causes directions in the eyepiece to be somewhat confusing until you are used to it. To get your bearings, turn off the telescope's motor-drive, if it has one, and let the image drift through the eyepiece field. Stars will enter from the east and depart to the west. Nudging the scope toward Polaris will indicate which direction is north.

There is another form of diagonal intended mainly for terrestrial use known as an *erecting prism*, which provides fully correct images. These typically turn the image 45 degrees instead of 90 degrees. Not only are they awkward to use for sky viewing, but the roof prism that erects the image produces an obvious line through bright objects viewed at night such as planets and 1st-magnitude stars.

Barlow Lenses

There is a wonderful little optical device that effectively doubles or triples the focal length of any telescope, yet it's only a few inches long! Called a *Barlow lens*, it consists of a negatively curved achromat (or sometimes three elements) fitted into a short tube, one end of which accepts the eyepiece while the other end goes into the telescope's focuser. With the preponderance of short-focus refractors and Dobsonian-style reflectors in wide use today, these "focal extenders" are enjoying renewed popularity among observers.

The Barlow's negative element decreases the angle of convergence of the light being brought to focus by the telescope's objective lens or primary mirror, causing the latter to appear to be at a much greater distance from the focus than it actually is. This effectively increases the original focal length/ratio. Most Barlows are made to amplify between two and three times ($2\times$ to $3\times$). Note that a Barlow's rated "power" is based on the eyepiece being placed at a set distance into the drawtube provided. The further the eyepiece is pulled back from the negative lens, the greater the amplification factor becomes. By adding extender tubes, some observers have pushed their Barlows to $6\times$ and more! Also, if placed *ahead* of a star diagonal instead of behind it as intended, the extra optical path length through the diagonal to the eyepiece will also greatly increase its effective power.

Lunar and planetary observers have long employed Barlow lenses as one of their standard tools of trade. So, too, have double star observers. While casual observing of wide pairs (those, say, 10 seconds of arc apart or more) is most pleasing at low magnifications with their accompanying wide fields of view, tighter doubles

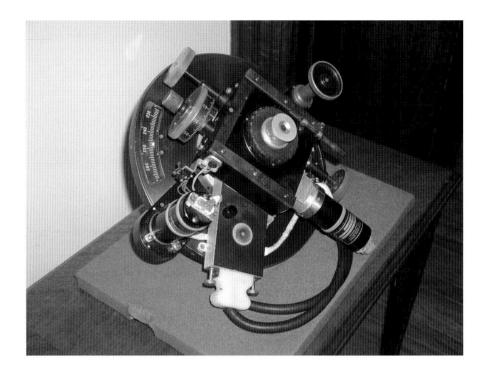

Figure 5.5. A filar micrometer was traditionally supplied with all research-grade (and many instructional-use) refracting telescopes a century ago, and these devices are still in use today. Seen here is the micrometer long employed on the U.S. Naval Observatory's 26-inch refractor (shown in Figure **5.2**) by the late astronomer Charles Worley to measure visual double stars. This instrument has several modern upgrades, including digitized readouts which eliminate the need to actually read the drumhead and position circle settings (for a pair's separation and position angle, respectively). Modern commercial filar micrometers are smaller, easier to use and more affordably priced for the amateur who wishes to measure double stars with this classical device. Courtesy U. S. Naval Observatory Library.

(especially those closer than 5 seconds) require higher powers to obtain a comfortable split. And for measuring binary stars (where separations are often under 1 second of arc), using the highest magnification the atmosphere will allow is the accepted rule. A Barlow lens makes it possible to obtain high powers using long focal length eyepieces – with their bigger lenses, often wider apparent fields of view, and more comfortable eye relief than short focal length ones.

Micrometers

A *micrometer* is an instrument for measuring very small distances or angles. It has been used in astronomy for determining the apparent dimensions of everything from sunspots and lunar craters to nebulae and galaxies. In its application to double stars, it is used to measure both the angular separation (in minutes, or more generally seconds, of arc) of two stars, as well as the position angle (in degrees and minutes) of the secondary star with respect to the primary or brighter one. Of the

Figure 5.6. A reticle-eyepiece micrometer assembled by American observer, Ronald Tanguay, and mounted on his 3.5-inch (90 mm) Maksutov-Cassegrain telescope. A protractor scale and pointer were added to a commercially available reticle guiding eyepiece, as seen in these two views. Using this simple device, he has made numerous double star measurements of high precision with this small-aperture instrument. Courtesy *Sky & Telescope*.

many types of such devices, we examine here a few of those most commonly used by double star observers today.

The traditional instrument employed for the measurement of doubles is the *filar* (or more accurately *bi-filar*) *micrometer*. This consists of a housing containing two very fine wires, one fixed in position and the other moveable on a precision screw connected to a graduated drumhead. Around the outside is a graduated position-angle circle. One end has a drawtube with an eyepiece in it and the other end has a drawtube that fits into the telescope's focuser.

Commercially made filar micrometers can still be found on the market today from a few select manufacturers. To find them, check the advertisements in *Sky & Telescope* and other astronomy magazines, or do a search on the Internet. Also, filar micrometers are often sitting idle in high school, college and university observatories (typically safely stored away in their scientific instrument rooms), just waiting to be put back into service by some dedicated observer. And a few observers have even successfully made their own units from readily available components.

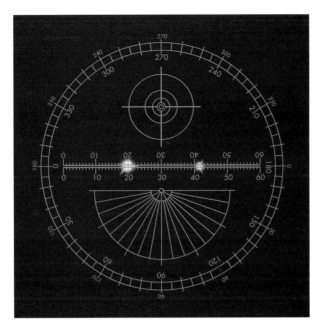

Figure 5.7. Once properly calibrated, a reticle eyepiece's linear scale provides an ideal means of determining the angular separation of a double star. In the hands of dedicated amateurs, these inexpensive oculars are transforming the serious observation of visual binaries traditionally measured using filar micrometers. Courtesy *Sky & Telescope*.

In use, the device is first turned so that the fixed wire bisects both stars at the same time; a pointer on the outer position circle indicates the position angle, generally read to the nearest 1/10th of a degree. This is measured from north (0 degrees), through east (90 degrees), to south (180 degrees) and finally west (270 degrees). The unit is then rotated 90 degrees, with the fixed wire this time bisecting only the primary star. The moveable wire is then used to bisect the companion star, and the angular separation is read from the drumhead – typically to an accuracy of 0.1 arc second or better. In practice, at least three bisections are made over several different nights and averaged for maximum accuracy.

However, before a filar or any type of micrometer can be used on the sky, it must first be calibrated for the given telescope–eyepiece combination. This is to determine the image scale their combined focal lengths/magnification give on the sky itself and is typically done by measuring several "standard fields" containing stars of known, fixed distances apart. References to the detailed calibration and use of this and other types of micrometers will be found in Chapter 6.

Another type of measuring device that has come into use on double stars in recent years is the *reticle eyepiece micrometer*. This employs one of several commercially available reticle or guiding eyepieces, which are widely used in taking photographs with a telescope. No wires or moving parts are involved – just an illuminated precision scale that, once calibrated, makes possible measuring the angular separation of stars to great accuracy. The only thing that needs to be added is a position-angle scale and pointer; a simple 360-degree protractor scale serves this purpose very effectively. References to several excellent *Sky & Telescope* articles on the use of this type of micrometer will be found in the next chapter.

Two other less common devices make use of multiple images for measuring double stars. One is the *double-image micrometer*, which employs optical wedges or prisms to produce two sets of images of a pair. The other is the *diffraction grating micrometer*, in which a series of evenly spaced slits or slats are placed over the front of the telescope to produce a graded series of images. By proper manipulation of the images created by these micrometers (superposing them on top of each other, for example), the angular separation of a pair can be determined, as well as its position angle. Even less used and known among amateur astronomers is the *chronometric micrometer*, which – as its name implies – times the transit, as determined by a stopwatch, of the primary and secondary star of a double across a fixed single wire in the eyepiece.

A number of high-tech methods for measuring close doubles have been devised in recent years by professional astronomers, one of the most successful and powerful of which is that of *speckle interferometry*. "Speckle" refers to the flickering interference/diffraction effect upon a star's image resulting from atmospheric undulations of the incoming wavefront. This pattern contains high resolution data, which can be extracted through the use of high-speed photography of very long focal length double star images made with observatory-class refractors. Using sophisticated optical (laser scanning of the photographed image) and mathematical (Fourier transform) procedures, it is possible to effectively "see" and resolve star images just 0.01 arc seconds apart.

Powerful as this technique is, it is effective only for double stars with a difference in brightness of no more than about three magnitudes, and it does not work for pairs wider than 4 or 5 seconds apart. This leaves a vast realm of opportunity for the dedicated amateur using more conventional devices like filar and reticle-eyepiece micrometers, as discussed in Chapter 6.

While on the topic of photographing double stars using long-focus refractors, it should be pointed out that this was the traditional method of measuring pairs wider than about 5 seconds apart, and it is still done at a few observatories today. While the author is almost exclusively a visual observer, he has participated in such work in the past using the huge 30-inch f/18 Thaw refractor (still the fifth largest in the world) at the University of Pittsburgh's Allegheny Observatory. Double star images taken on precision 8"×10" glass photographic plates were then measured for separation and position angle (as well as parallax) using state-of-the-art measuring machines. To the author's knowledge, no amateur is presently using either long-focus photography or speckle interferometry – both of which require costly equipment and specialized training – to measure double stars.

A relatively recent development in the measurement of double stars is the use of video and CCD imaging, which are discussed in the next section. In its simplest form, the image of a pair is electronically amplified and displayed on a television or computer monitor, where the separation and position angle are then measured directly off the screen itself against a displayed grid-work of known angular dimensions. The next chapter references an excellent article from *Sky & Telescope* explaining this promising technique in depth.

Photographic, Video and CCD Imaging Systems

In a feature article on observing double stars in the March 1993 issue of *Sky & Telescope* (and several years earlier in both *Astronomy* and *Star & Sky* magazines), the author first suggested that astrophotographers attempt to capture the hues of

Figure 5.8. The video-imaging setup of German amateur Rainer Anton on his homemade 10-inch Newtonian reflector. An inexpensive commercial black and white video camera is used to shoot tricolor images through red, green and blue filters. The camera head and filter wheel is seen here attached to the telescope's 2-inch focuser. A camera controller and digital video recorder are seen in the foreground. Courtesy Rainer Anton.

some of the sky's brighter and more colorful pairs. As an aside in that article, I mentioned that video and CCD imaging could also possibly be used for this purpose. These suggestions have certainly borne fruit, as many observers today are routinely imaging these formerly neglected jewels of the heavens.

Over the years, celestial photography has been done with everything ranging from basic 35mm cameras (either tripod-mounted or piggybacked onto telescopes) and telescopic astrocameras to sophisticated astrographs and Schmidt cameras. But the use of film to capture the sky by both amateur and professional astronomers is rapidly disappearing in favor of electronic imaging.

At the forefront of this revolution is the widespread use of *charged coupled devices*, or *CCDs*. These detectors use a photon-sensitive silicon chip – or often a mosaic of them – placed at the telescope's focus. There, each incoming photon is essentially multiplied many times, and the resulting energy from the chip is transmitted in the form of an electrical signal to a monitor for viewing and/or electronic recording.

As noted in the next chapter, one of the difficulties of attempting to capture the hues of double stars on color film (unless the stars are of nearly equal magnitudes) is that exposing long enough to image the fainter companion overexposes or "burns out" the brighter primary; or if the exposure is just right to record the color of the primary, the companion is underexposed. To be sure, there are ways around this (using two separate exposures and then combining images, or using an occulting strip in the eyepiece to hide the primary during part of the exposure), but they require skill, patience, and generally lots of exposures before finally capturing a pair in its natural color.

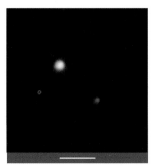 

Figure 5.9. Two video images obtained with the equipment shown in **Fig. 5.8.** On the left is the well-known visual double star Cor Caroli (α Canum Venaticorum), a lovely showpiece in the smallest of telescopes. On the right is Herschel's Wonder Star (β Monocerotis), one of the finest visual triple systems in the sky and a marvelous sight in medium-aperture amateur instruments. Courtesy Rainer Anton.

CCD detectors not only have an immensely faster speed (or high "quantum efficiency") than do film emulsions, but they also have an amazing "dynamic range". This means that a bright star and faint star right next to each other will both be properly "exposed" on its chip. Their high speed and corresponding short exposure time per image also mean that they can capture those fleeting moments of steady seeing, resulting in sharper images of close doubles than can be achieved with conventional film.

Another exciting related development in imaging has been the commercial introduction of inexpensive *eyepiece video cameras* that can be used on virtually any telescope to rapidly capture pictures of celestial objects in either black and white or color and can display them on a conventional television screen. Many observers are also adapting common *web cam* video cameras to their telescopes for video imaging (especially of the Moon and planets), often with surprisingly good results.

Star Atlases and Catalogues

An explorer of the heavens without a good star atlas is like a tourist without a road map – both wandering aimlessly without any idea of where they are. Of the many celestial maps, atlases and catalogues available to the stargazer today, we concentrate here on those of particular interest and use to the double star observer.

The author's personal favorite atlas for such work is the "old" *Norton's Star Atlas* – any one of the 17 editions of this classic *before* the appearance of the newer 18th edition in 1989, renamed *Norton's 2000.0*. While the latter now has a symbol for double stars (a line bisecting the star-dot) which the former did not, its editors have sadly dropped the priceless double star designations that were a hallmark of the earlier editions. These consisted of the official symbol or abbreviation for the discoverer of each pair plotted and the catalog number from the corresponding catalogue. Both the old and new editions have the wonderful advantage over more detailed atlases of showing nearly the entire sky visible at a given time of year on a double page, except for the polar regions, which are on separate maps. Both works also contain listings accompanying the maps of some of the more interesting double stars and other deep-sky objects plotted.

In the author's opinion, the finest all-purpose star atlas in existence is Wil Tirion's superb *Sky Atlas 2000.0*, co-issued by Sky Publishing Corporation and Cambridge University Press. Tirion is generally recognized as the greatest celestial cartographer in the world today. This work is a joy to look at (especially the deluxe color-coded edition) and exciting to use at the telescope, plotting over 81,000 stars to visual magnitude 8.5 along with some 2,700 clusters, nebulae and galaxies over the entire visible heavens on 26 large-scale charts.

A two-volume catalog entitled *Sky Catalogue 2000.0*, by Alan Hirshfeld and Roger Sinnott, provides data for most of the objects plotted in the atlas. Arranged by object type, the double and multiple star section contains over 8,000 entries. (Also available is the more recent *Sky Atlas 2000.0 Companion* by Sinnott and Robert Strong, which contains descriptions of every nonstellar object plotted on the atlas.)

Other much more detailed atlases have appeared in recent years, such as *Uranometria 2000.0* and *The Millennium Star Atlas*. Plotting hundreds of thousands (or more!) stars and deep-sky objects, they were published partly in response to the growing use of truly huge, large-aperture Dobsonian-type reflectors by amateur astronomers today. In practice, these massive volumes are often unwieldy and confusing to use at the eyepiece at night. For the purposes of most double star observers, they are definitely overkill.

The ultimate reference in this field is the *Washington Double Star Catalog*, or *WDS*. Compiled and maintained by the United States Naval Observatory in Washington, D.C., it is the world's largest collection of data on these objects. Incorporating all previous great discovery catalogues, together with more recent finds and the latest measures of pairs from observers around the world, the latest edition contains entries for over 98,000 double and multiple star systems. It can be accessed on-line at the Observatory's Internet site at http://ad.usno.navy.mil/wds/ (which contains a veritable galaxy of information about these objects) and is also available without charge on a CD-ROM. In addition, a valuable catalog of current visual binary orbits can also be accessed on-line or requested on CD-ROM. Further information is provided on the site itself, where both CDs can be ordered on-line. Those without a computer can contact the Observatory directly by mail at Massachusetts Avenue, NW, Washington, DC 20392, USA.

An extensive listing of official double star designations and catalogues is given in Appendix 2. The symbols and abbreviations shown are based upon the names of the various discoverers, measurers or their institutions, and they span the period from some of the very earliest to the very latest such compilations.

Miscellaneous Items

In this final section, we touch briefly on a number of other stargazing tools in addition to the major ones already discussed. The observer's specific interests and needs will determine how useful they will be.

Binocular Eyepiece

This device makes possible the relaxed use of both eyes with refracting and catadioptric telescopes (reflectors generally do not have enough focus travel to allow its use). The unit itself is rather expensive, and it requires two eyepieces having identical focal lengths. While providing spectacular 3-D views of the Moon, planets and brighter deep-sky wonders, it has found little use among double star fans.

Dew/Light Shield

While reflectors have their own built-in version of these, as their primary mirrors are placed at the bottom of their tubes, refractors and catadioptric telescopes need these extensions to their tubes. They both prevent dew from forming on their up-front optical elements and also shield stray light from entering the telescope. Refractors generally are equipped with a dew shield, although most are much too short to be of real value. And surprisingly, virtually all manufactured catadioptric telescopes come without one at all! In either case, the observer can fashion one out of some opaque, black flexible material like common poster board or can purchase one commercially.

Heated Eyepieces

This related item consists of an electrically heated strip attached to the eyepiece (and sometimes also to a telescope's objective lens or corrector plate) to raise its temperature just enough above the ambient to prevent the formation of dew. Again, both homemade and commercial units are seen on many telescopes today.

Photographer's Cloth

This is simply a dark, opaque cloth that is thrown over the observer's head (generally down to waist level) to prevent stray light from destroying the eye's dark adaptation. In practice, it can be a bit suffocating, especially on muggy nights, and it's sure to raise the eyebrows of any neighbor who happens to see you lurking in the dark.

Aperture Masks

When atmospheric seeing is less than ideal, observers have found that reducing a telescope's aperture (especially in the case of those 12 inches or larger) can improve image quality and image motion. These masks are typically made by cutting an opening in a piece of cardboard smaller than the original aperture and placing it over the front of the telescope. Note that the opening can be on-axis for a refractor, but should be placed off-axis for reflectors and catadioptrics to avoid their central secondary mirror obstructions (which limits the mask's clear aperture to less than the radius of the primary mirror). While this often does improve image quality on the Moon, planets and bright stars, reducing a telescope's aperture also reduces its resolution to that of the size of the opening in the mask.

Filters

Color filters have long been used by planetary observers to enhance certain features, and many deep-sky observers today routinely use light pollution and nebula filters to increase the visibility of faint objects. While a few amateurs have used color filters in an attempt to determine the visual tints of doubles stars, filters in general offer little benefit in viewing these objects due to their intrinsic brightness and the fact that they are essentially point sources.

Setting Circles/Go-To and GPS Systems

The use of traditional mechanical setting circles on the telescope displaying Right Ascension and Declination in the sky to find celestial objects is rapidly disappearing in favor of the much easier to use and more precise digital circles, computerized "go-to" systems, and the truly amazing GPS or Global Positioning Satellite systems. For us purists, these devices take much of the fun out of celestial exploration and leave their users not knowing the sky without their aid. In any case, most double stars of interest to amateur astronomers are naked-eye objects that can be easily sighted directly or found by the traditional technique of "star-hopping" from bright stars to the target of interest using a star atlas.

Mountings and Motor Drives

The two basic types of telescope mounting are the simple *altazimuth* and the more complex *equatorial*. The former provides basic up-and-down (altitude) and around (azimuth) motion on the sky, and is the principal behind the immensely popular Dobsonian reflecting telescope. Altazimuths are ideal for casual stargazing. The equatorial mount is the more costly and heavier (most forms requiring counterweights to balance the weight of the telescope itself) of the two, but has the important advantage that it can be equipped with a motor (or "clock") drive to counter the diurnal rotation of the Earth on its axis that causes objects to slowly drift out of the field of view. This feature is essential for observing double stars at high powers, especially when undertaking micrometer measurements of close pairs. With the advent of computerized drives that track in two coordinates, it is now possible to retain the simplicity and lower weight of the altazimuth mounting for such work.

Computers

It goes almost without saying that computers have become an important tool in observational astronomy, as in almost every other area of modern life. They are used for such tasks as aiding telescopes in finding and tracking celestial objects (often remotely, as from the observer's living room, den or office), and in making, processing and displaying observations by video and CCD imaging.

Observatories

Fortunate indeed is the observer who has a proper shelter for his or her telescope – one where the instrument can be left safely outdoors, protected from the elements, and be ready to use almost immediately upon demand. The *domed observatory* is the best known and most aesthetic of such structures and offers the maximum protection from wind and stray lights. But it is also the most costly, requires the longest cool-down time and provides only a limited vista of the sky through its narrow slit. An alternative to the dome is the *roll-off roof observatory*, in which the entire top of the structure rolls back on tracks to reveal the whole visible heavens. Depending on the height of its walls, it may offer only limited protection from wind and lights. A delightful compromise between these two types is the *flip-top roof observatory*. Here, a structure with low walls and hinged peak roof splits into two sections, one half typically swinging to the east, the other half to the west. Chains or ropes control how high or low the halves extend. The observer adjusts them to reveal whatever part of the sky is being viewed, and at the same time uses them as a shield against wind and lights. Most observers of the past – including the greatest observational astronomer of them all, Sir William Herschel – worked in the open night air unprotected by any kind of structure. So, too, do a majority of stargazers today, including the author (who has used all three types of observatories over the years as both an amateur and professional astronomer).

Chapter 6

Observing Projects

In this chapter, we examine seven observing projects for double star enthusiasts. These range from very basic and easy to quite advanced and involved. In each case, be sure to study the guidelines discussed in Chapter 4 on training the eye and the various observing techniques before embarking on them. Above all, approach these projects with a sense of sheer cosmic fun and deep personal enrichment rather than as serious undertakings!

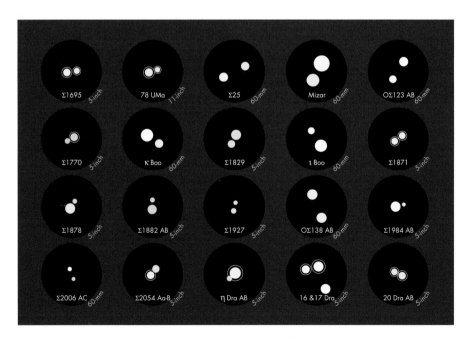

Figure 6.1. Double star drawings by Sissy Haas using 2.4-inch and 5-inch refractors (and for one of them an 11-inch refractor). Unlike her low-power sketches in **Fig. 1.2.**, these were made with relatively high magnifications in order to show the diffraction disks of the components. According to many observers, this makes determining the tints of the individual stars much easier than if they were essentially point sources, as is the case with low powers. Courtesy Sissy Haas.

Sightseeing Tour

The very first thing any observer should do is to undertake a survey of the sky's brightest and finest double stars. The descriptive roster of showpieces in Chapter 7 is an excellent place to begin. Other more extensive listings are offered in such delightful classic works as *Celestial Objects for Common Telescopes* by T. W. Webb (Dover), *Field Book of the Skies* by W. T. Olcott (Putnam) and *New Handbook of the Heavens* by H. J. Bernhard, D. A. Bennett and H. S. Rice (McGraw-Hill). Unfortunately, these are currently all out of print, but are still to be found in many libraries and used book stores.

The author's own 1998 self-published work, *Celestial Harvest: 300-Plus Showpieces of the Heavens for Telescope Viewing & Contemplation* (a 40-year labor of love!) was reprinted by Dover Publications in New York in 2002 and is now widely available. Nearly half of its entries are spectacular double and multiple stars suitable for viewing with backyard instruments. Dover is also the publisher of *Burnham's Celestial Handbook* by the late Robert Burnham, Jr., which contains data for thousands of visual pairs and detailed descriptions for many of the brighter and better known ones.

Bright doubles abound among the naked-eye stars. They are, therefore, easily found and observed, even on nights that are useless for viewing other types of deep-sky wonders due to haze, moonlight or light pollution. Their endless profusion and variety of color, magnitude, separation and component configuration make these objects fascinating even to seasoned double star aficionados. Literally thousands of these tinted gems, no two of which are exactly alike, lie within reach of the smallest glass.

So make your first priority as a double and multiple star observer to become familiar with as many of these lovely objects over the course of the viewing year as possible. Browsing through any of the above listings is sure to motivate you to not only seek out additional specimens, but also to become more deeply involved in observing those that you have already become acquainted with.

Color Studies

An ideal follow-up project to (or one concurrent with) the initial sightseeing tour involves the fascinating as well as controversial subject of double star colors. The hues assigned to these objects are often widely discordant, and in some cases even bizarre. Involved are such complex and variable factors as telescope optics, atmospheric conditions, color contrast effects, magnitude difference and spectral type of the components, as well as the observer's color perception and "preconditioning" from reading published descriptions. Thus, in addition to the conventional pure colors of the spectrum, we also find reported such tints as emerald, turquoise, rose, olive, copper, topaz, garnet, ashen, primrose, scarlet, gold, pearl, ruby, purple, diamond, sapphire, lavender, lilac, grape, tiger-lily and even beryl-sardonyx!

No systematic effort has been made in recent times to survey and record the colors of even the brightest pairs, nor to faithfully reproduce them for others to see. Yet these heavenly hues are there in the sky to be seen, illusory or not. As J.D. Steele pointed out long ago, "Every tint that blooms in the flowers of summer, flames out

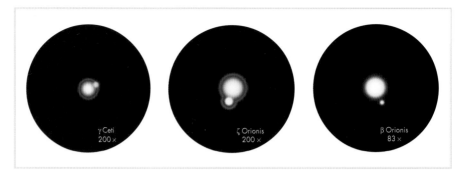

Figure 6.2. Three additional drawings of double stars by Sissy Haas, using her 5-inch refractor. The eyepiece sketches of γ Ceti and ζ Orionis were made at 200×, and show obvious stellar diffraction disks and rings. Although these features are still present in the lower power view of β Orionis (Rigel) made at 83×, they are smaller and much less obvious. The extended image size of the primary in this case is due mainly to its overpowering brilliance, shining at a visual magnitude of 0.1. Courtesy Sissy Haas.

in the stars at night." This is especially obvious when two or more radiant suns are in close proximity to each other within the eyepiece field.

It should be mentioned here for the record that many of the contrasting double star colors actually *are* real, resulting from differences in spectral class (and, therefore, temperature) of the individual suns and not the result of contrast effects. This has been pointed out by many of the classic visual observers. One famous example involves the beautiful reddish-orange supergiant Antares and its emerald-green companion. During an occultation of the pair by the Moon, the noted double star observer William Dawes witnessed "A curious proof of its independent, not contrasted green light, when it emerged, in 1856, from behind the dark limb of the Moon." You can see this for yourself anytime by positioning one of the components of a bright tinted double like the well-known orange and blue pair Albireo (β Cygni) behind a foil strip placed across the field stop of your telescope's eyepiece. Hiding either star makes no difference in the other's perceived tint.

In the case of color studies, telescopes of 6-inch aperture and less actually have an advantage over larger ones, since small instruments provide an optimum light level for viewing the hues of the brighter and more beautiful doubles. Too much illumination (as well as too little) makes color perception difficult and uncertain. Over the years, the author has worked with telescopes of up to 36-inch aperture (including a 30-inch refractor!), and has never found the tints of such dazzling pairs as γ Andromedae, ε Bootis, α Herculis, γ Delphini or lovely Albireo itself as vivid in these big scopes as in a 3- or 4-inch glass.

Furthermore, the smaller the telescope aperture, the larger the spurious stellar diffraction disk that will be seen. A 2- or 3-inch instrument at magnifications of 100× or more shows a very sizable disk, making even subtle hues easier to detect than looking at a point-like stellar image. Observers having larger instruments can achieve the same effect by placing a cardboard aperture mask with an opening of this size over the front of the telescope; this should be done on-axis in the case of refractors, and off-axis with reflectors and compound scopes (thereby avoiding their central obstructions). The resulting transformation in apparent image size, and with it color enhancement, for the brighter double and multiple stars (and for the brighter single stars as well) is truly astounding.

Observations are best made at one or two fixed magnifications using high-quality, well-corrected eyepieces such as orthoscopics or Plossls, and with the stars centered in the field of view to minimize color aberrations within the optics. As haze, bright moonlight and auroras are known to affect color perception, making color estimates should ideally be reserved for dark, transparent nights. The target pairs should also be on or near the meridian – and, therefore, at their highest altitude in the sky – to minimize the effects of atmospheric turbulence and absorption. Finally, in viewing fainter types of deep-sky objects, averted vision is typically used to bring the eye's light-sensitive outer rods into play. However, determining the tints of doubles should be done by staring directly at the stars, which activates the color-sensitive cones at the center of the eye.

A set of perhaps a dozen "standard colors" should be adopted in logging the tints seen. Such a tabulation of star colors, all made by the same observer using the same telescope and eyepiece combination under good sky conditions, would be of real value and interest to other amateurs. Comparing the visual colors seen with the stars' published spectral types would be another fascinating application of the results. Going a step further, observers with an artistic bent might try sketching some of the more colorful pairs in pastels or other suitable medium.

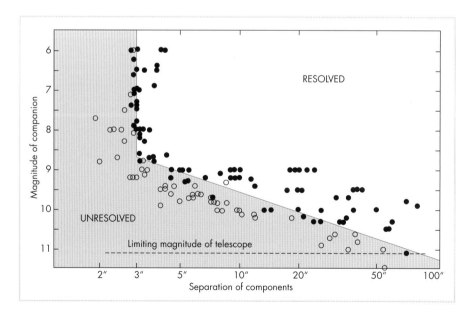

Figure 6.3. A Peterson diagram, originally compiled by American amateur Harold Peterson half a century ago using a 3-inch refractor at a fixed magnification of 45×. It shows that resolution of a pair is independent of the companion star's brightness until about 9th magnitude (for a 3-inch aperture), beyond which fainter stars require greater separations to be seen. Since 45× is significantly below the recommended telescopic "resolving magnification" of at least 25× per inch of aperture (or 75× for a 3-inch glass), double star observers desiring to create a Peterson diagram for their telescope should use higher powers than that employed in the original study.

More challenging, astrophotographers might attempt capturing these lovely gems on film. There are difficulties, to be sure. In shooting pairs of unequal brightness, for example, exposing long enough to capture the companion's hue often overexposes the primary's image and turns it white no matter what its true color. Furthermore, the color recorded by films sometimes depends on the length of the exposure; green overexposing to yellow is one common example.

Perhaps the most exciting activity here for those observers equipped with eyepiece video cameras and CCD imaging systems is recording double star colors electronically. The excellent color sensitivity/response of the latest such devices on the market, coupled with their high dynamic range (i.e., ability to image bright and faint objects equally well), make them ideal for rapidly shooting and processing lots of pictures in real time. Such ultra-short exposure times insure that at least some of the images will capture those relatively rare moments of superb seeing that make for razor-sharp pictures.

It is to be hoped that someday soon astronomy magazines, observing guides and textbooks will finally be graced with actual color images of these jewels as astrophotographers and astroimagers act on these suggestions. This area of activity is wide open to anyone who does so.

Resolution Studies

Another challenging but potentially rewarding project for observers is a much-needed review and revision of the famous empirical relationship for resolving double stars devised by W.R. Dawes of England in the late 1800s. He found from observations with several excellent refractors of various sizes that $R = 4.56/A$, where R is the resolution in arc seconds and A is the telescope's aperture in inches. (If expressing the aperture of the telescope in millimeters rather than inches, the relationship becomes $R = 116/D$), where D for diameter has the same meaning as A. But this formula holds strictly true only for pairs of equal brightness and of about the 6th magnitude. For brighter, fainter and especially unequal pairs, Dawes' Limit departs markedly from actual results at the telescope (values as great as $36/A$ having been reported in the case of a 6-magnitude difference between a primary and its companion).

Another resolution relationship is the *Rayleigh Criterion*, formulated by the British physicist Lord Rayleigh. Here $R = 5.5/D$, where R is again the resolution in arc seconds and D is the diameter (or aperture) of the telescope in inches. This theoretical relationship is based on the wave nature of light and give a somewhat less stringent and (according to many observers) more realistic result. Part of the difference involves what is actually meant by "resolution" of two stars. Dawes' Limit considers a pair resolved when the first dark ring of one star's diffraction pattern intersects the center of the other star's central disk – which means a notched or partially merged image of the pair. The Rayleigh Criterion considers a pair split when the outer edge of each star's diffraction disk is separated by a space equal to the width of the first dark ring – in others words, separated images. Professional observers like William Markowitz of the U.S. Naval Observatory have found that for a pair to show disks just in contact, $R = 6/D$ gives a more realistic value of results at the eyepiece. This is known as the *Markowitz Limit*.

The primary factor at play here seems to be the magnitude difference between components. The need for adding some form of "magnitude term" to the above

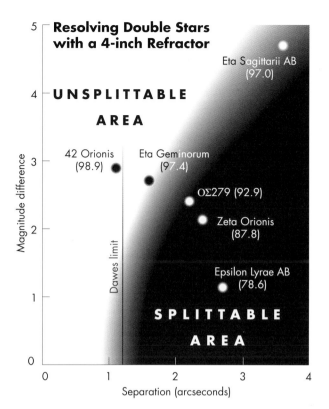

Figure 6.4. This diagram, compiled by Spanish observer Luis Arguelles for his 4-inch refractor, shows that resolving close double stars is not always a simple matter of "it is" or "it isn't." Many factors are involved, including the observer's experience. He has developed a computerized "Difficulty Index" (or DI) utility for double stars, with values ranging from 0 (very wide, easily split pairs) to 100 (extremely difficult ones). The DI is indicated in parentheses for those doubles plotted on the graph. Courtesy *Sky & Telescope*.

Figure 6.5. Luis Arguelles seated at the 4-inch refractor used for his double star work, which includes development of the Difficulty Index discussed in **Fig. 6.4.** He has done much to foster both casual and serious observing of these fascinating objects by amateurs worldwide. Courtesy Luis Arguelles.

three relationships seems clear, but until recently little has been done by double star observers towards determining one. Derivation of such a term could well result from a long series of carefully made observations of many pairs. The extended listing of 400 double and multiple stars in Appendix 3 offers an excellent source of targets for such a study. A more exhaustive one can be found in Volume II of *Sky Catalogue 2000.0* by A. Hirshfeld and R. Sinnott (Sky Publishing), which lists more than 8,000 visual double and multiple stars over the entire sky.

Observations are best made on several nights to average out atmospheric and other effects. A standard single magnification should be used, one equal to or greater than the "resolving magnification" of the telescope. This is usually given as 25× per inch of aperture (200× on an 8-inch, for example), but some authorities quote values as great as 40× per inch. The author feels this is too high and that a good compromise is perhaps 30× per inch.

Good atmospheric seeing (image steadiness) is a must, and often occurs on hazy or even muggy nights. The telescope optics need to be of excellent quality and well collimated, as well as clean and free of light-scattering dust. Although accurate visual magnitudes and current angular separations for each pair will be required for the final analysis, they do not need to be known for actually making the observations (and, in fact, it will help *not* to know them to avoid bias at the eyepiece). It is essential, however, to log the date of each observation, the seeing conditions, and most importantly whether the pair is resolved (with dark sky between the stars), blended (star image notched or egg-shaped), or not resolved (image appears as a single star only). You will eventually end up with a list of double stars that are and are not resolved for a range of magnitude differences for your particular telescope. This can provide the statistical data required for you – or perhaps someone more mathematically inclined – to arrive at the desired term.

Interestingly, here again small scopes have an advantage over larger ones. Apertures of 4 or 5 inches are virtually independent of all but the worst seeing conditions. They can often split such well-known test objects as ε Bootis, ζ and η Orionis, and ε Lyrae (the "Double-Double") on nights when observatory-class instruments show only fuzzy flickering blobs of light.

One fascinating early attempt at arriving empiricially at a better double star resolution criterion was conceived and undertaken by American amateur Harold Peterson and has appeared in *Sky & Telescope* magazine many times over the years. Widely known as the "Peterson diagram," his approach was to plot the resolution and nonresolution of a selected list of double stars against the apparent magnitudes of their companions, as seen through his 3-inch refractor. From these observations, he formulated a prediction of the faintest magnitude companion that could be resolved with a given telescope at a given separation.

Peterson found that splitting a pair was actually independent of the companion star's brightness for his 3-inch aperture down to about 9th magnitude, beyond which fainter stars required greater separation to be resolved. In particular, his results indicated that the main factor affecting double star resolution was not the actual magnitude difference between the components of a pair, but rather the difference in magnitude between the fainter star and the magnitude limit of the telescope. This work should definitely be extended to other sizes and types of instruments used by today's amateur astronomers, and should be conducted over a much larger sample and wider range of doubles, bringing with it valuable insight into the specific nature of the sought-after magnitude term.

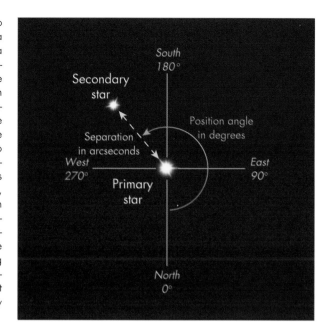

Figure 6.6. The two primary pieces of data obtained in measuring a visual binary at the telescope (in addition to the date of the observation itself) are the angular separation of the pair and the position angle of the companion with respect to the primary (usually brighter) star. North in the sky is shown here at the bottom, matching the view through an ordinary inverting telescope; use of a star diagonal results in a mirror image of this diagram, having north at the top of the eyepiece field, and with east and west reversed. Courtesy *Sky & Telescope*.

Two relatively recent contributions to this subject appeared in the very same issue of *Sky & Telescope* for January, 2002. One is a monograph by mathematician Christopher Lord from England that takes into account the difference in brightness of a pair's components and provides a "performance index" for a given size telescope. Spanish observers and computer experts Luis Arguelles and Rafael Barbera have developed a freeware utility that will provide a "difficulty index" for any pair listed in various catalogs, including the massive *Washington Double Star Catalog*'s some 98,000 entries, in just a matter of seconds.

Yet, to be truly definitive, any revision of Dawes' Limit will require observations to be made over as great an aperture range as possible, and must take into account the results from not only refractors, but also from reflectors and compound catadioptric instruments (whose central obstructions are known to impact resolution as well as image contrast). The work is great and the challenges are real but the observer who comes up with the correct magnitude term will be immortalized.

Micrometer Measurements

The discovery and measurement of binary stars – those actually orbiting each other and not merely aligned along our line of sight or drifting through space together as common proper motion pairs – was a high priority among both amateur and professional astronomers up to the early part of the last century. The field was alive with the likes of the Herschels, the Struves, Dawes, Burnham, Aiken and Barnard. But with only a handful of observers active today, the task of tracking the orbital motions of the more than 98,000 cataloged pairs is apparently hopeless. Yet this brings with it an opportunity for the dedicated, well-equipped amateur to instantly join ranks with the few remaining professionals still engaged in such research.

Figure 6.7. With these two simple, commercially available tools any amateur having a steady, equatorially mounted telescope with good optics can do valuable double star work. Shown is an illuminated reticle eyepiece for making the actual measurements and an ordinary scientific calculator for reducing the observations. Another reticle eyepiece micrometer appears as **Fig. 5.6** in the previous chapter. Courtesy *Sky & Telescope*.

This work involves the routine measurement of generally close binaries, especially at crucial times in their elliptical orbits. Three basic parameters are obtained in such observations. One is the *position angle* of the companion with respect to the primary star. This is measured from north (0 degrees), through east (90 degrees), to south (180 degrees) and finally west (270 degrees). Another is the *angular separation* of the components in seconds of arc. The former is typically measured to the nearest tenth of a degree and the latter to an accuracy of 0.1 arc second or better. The final parameter is *time* – the date of the observation, designated by the year and decimal fraction thereof.

Since the orbital periods of most close binaries visible in amateur instruments span many decades or even centuries, position angle and separation vary quite slowly. However, when near periastron (at which time the stars are physically

Figure 6.8. British amateur Thomas Teague, seen here seated beside his 8.5-inch Newtonian reflector, has done some excellent double star work using an illuminated reticle eyepiece like that shown in **Fig. 6.7.**, thus dispelling the myth that you can't make useful visual measurements with a reflector. His technique eliminates the need for a position circle to find the position angle of a pair. Courtesy Thomas Teague.

closest together), these parameters can change quite rapidly and are spectacular to follow. A celebrated case is the 3rd-magnitude pair Porrima (γ Virginis). When at its widest separation of 6 seconds (which last occurred in 1919), it is an easy object in even the smallest glass. Although closing up for many decades now, it can still be split in backyard telescopes. But by the time of its next periastron passage in 2005, it will be unresolvable in all but the largest instruments. The pair (then separated by just 0.4 seconds) will appear as a merged elliptical blob with a rotating major axis – the stars flying around each other at an angular rate of more than 70 degrees per year!

The traditional instrument for measuring visual binaries is the *filar micrometer*, which is described in detail along with other double star measuring devices in Chapter 5. A filar micrometer's construction and operation are described in such classic works as R.G. Aikens's *The Binary Stars* and P. Couteau's *Observing Visual Double Stars*. A more recent useful reference is Volume I of the Webb Society's *Deep-Sky Observer's Handbook* edited by K.G. Jones. The superb two-volume set *Amateur Astronomer's Handbook* and *Observational Astronomy for Amateurs* by J.B. Sidgwick, published three decades ago and reprinted by Dover in 1980, contains very valuable information about binary stars, including their measurement, orbit calculation and plotting procedures.

Another valuable reference for those seriously thinking about entering this area of double star work is the late Charles Worley's 1961 reprint (updated in 1970) *Visual Observing of Double Stars* from his acclaimed *Sky & Telescope* series of the same title. The section entitled "The Measurement of Visual Double Stars" is especially useful and has a carefully selected working list of pairs for study (the data for which itself is largely out of date). Unfortunately, this little booklet has been out of print for some time. Through the kind permission of Sky Publishing Corporation, this section has been excerpted and appears in this volume as Appendix 5.

In 2004, Springer-Verlag released the book *Observing and Measuring Visual Double Stars* edited by Bob Argyle, president of the Webb society. This new work provides in-depth information on all aspects of measuring doubles using a variety of instruments, including reticle eyepiece and filar micrometers.

Special mention must be made here of *The Double Star Observer's Handbook* by Ronald Tanguay, whose work in measuring double stars and reviving interest in this neglected field is discussed below. This superb reference covers all aspects of visual double star observing and is, in this author's opinion, the most detailed and complete reference available today for anyone seriously interested in this field. It provides observers with a worthy successor to Aitken's out-of-print classic *The Binary Stars*. For more information, Tanguay can be contacted at the website and postal address given in the section at the end of this chapter on founding a double star observers' society.

Three articles in *Sky & Telescope* in recent years deserve special mention here. In the February 1999 issue American observer Ronald Tanguay describes in "Observing Double Stars for Fun and Science" his method of using a modified reticle guiding eyepiece to make professional-quality measurements using a small Maksutov–Cassegrain telescope – one just 3.5 inches in aperture! The other two articles both appeared in the July 2000 issue. One, entitled "Double-Star Measurement Made Easy" by British observer Thomas Teague, also describes making measurements with the same device, while "Measuring Double Stars with Video" by German amateur Rainer Anton explores the use of CD imaging to measure pairs. All three are superb modern contributions to the subject and should be read by anyone contemplating entering this field. (Please note that all of the *Sky & Telescope* articles referenced above can be accessed on-line at skyandtelescope.com.)

The requirements of a telescope intended for this type of work are few but demanding. It must have good optics and a rock-steady, motor-driven mounting, since observations are made at the highest magnification the atmosphere will allow (typically 50× to 100× or more per inch of aperture). Measuring star images is simply impossible if the instrument drifts or shakes. The minimum aperture required is largely a matter of opinion and primarily depends on the skill of the observer. Burnham, one of America's most famous double star observers, did much of his early work with a 6-inch refractor. Telescopes with long focal ratios (f/10 or higher) are preferred over short-focus ones due to their greater image scales. Even when using a refractor (the traditional instrument of choice for such work), double star observers generally employ a Barlow lens to achieve the maximum possible image size.

Long-focus Newtonian reflectors have been successfully used by some observers for measuring doubles, as have Maksutov- and Schmidt-Cassegrain telescopes.

However the latter present a problem; the most popular designs are focused by moving the primary mirror, which in turn causes the image scale to vary as the focus is changed. It is unclear if micrometer observations would be compromised, but repeated measures of standard pairs of known separation could settle the question. On the positive side, some observers claim that the relatively large secondary obstructions of compound systems like these actually improve the resolution of bright, nearly equal pairs by favorably altering the diffraction pattern. They also feel this makes measurement easier, since the enlarged spurious star-disks are less difficult to bisect than are point images.

Unlike color studies, measuring binary stars requires good seeing much more than high sky transparency. Hazy, muggy, moonlit, and even light-polluted skies that are unfit for other types of deep-sky observing often prove superb for measuring binaries. Stagnant air indicates a tranquil atmosphere with little or no turbulence, resulting in sharp, steady star images.

For the purposes of the visual double star observer seeking to draw up an observing program, these objects may be divided into two classes: fast-moving pairs and neglected pairs. The fast-movers are of the most interest since they have relatively short orbital periods, but they also require large apertures to measure. They are also the binaries that active professionals go after with large observatory refractors, using advanced measuring techniques such as speckle interferometry. This interference-based apparatus provides sub–arc second accuracy for very close pairs, but becomes ineffective on wider ones – those separated by 5 seconds of arc or more. This falls nicely within range of pairs that can be readily measured by the amateur astronomer with a filar or reticle eyepiece micrometer.

It is this area of wider, neglected pairs where the amateur has an unlimited opportunity to contribute valuable observations to the professional. There are thousands of such doubles within reach of backyard telescopes going unobserved, some of which have had no measurements made of them in more than a century! The best way to draw up an observing list of prospective candidates is to spend cloudy nights studying the *Washington Double Star Catalog* – either directly on-line at the United States Naval Observatory's web site at http://ad.usno.navy.mil/wds/ or on CD-ROM (which can be ordered without charge on the site itself). This continually updated reference source shows among other things the most recent date for which measurements have been reported for any given pair. It can be used to select neglected objects that are visible from your observing latitude, and that lie within reach of your telescope's magnitude and resolution limits. The site also offers its own lists of neglected pairs needing observations, compiled by Declination (northern, equatorial and southern lists).

Despite the long-term nature of double star measurement, there exists an urgency about such work. If binaries go unobserved at critical points in their orbits, it may well be centuries before we can catch them there again. Not surprisingly, the traditional rally-cry among double star observers has long been "Remember – the stars will not stand still!"

Orbit Calculation and Plotting

While not strictly an observing project in itself, the calculation of the orbital elements and plotting of orbital paths of double stars is both the direct result of

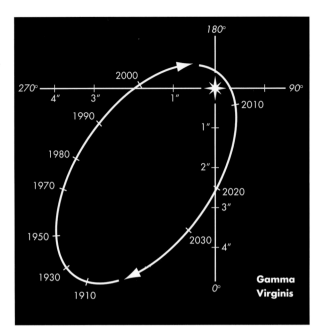

Figure 6.9. Orbit of the well-known visual binary γ Virginis (Porrima). Having a period of 171 years, this pair is closing to its next minimum separation (periastron) of 0.4 seconds in 2005. Note its rapid angular motion around that time, as compared with that at the opposite end of its circuit. Courtesy *Sky & Telescope.*

and ultimate purpose for observations made at the telescope. And it is the orbits of double stars that provide our only direct method for determining stellar masses, which are vital to our understanding of the structure and evolution of stars in general. These masses are derived from the observationally determined elements using Kepler's harmonic law of orbital motion.

Four primary "dynamic" elements sought after are the *semimajor axis* of the true orbital ellipse (*a*) in arc seconds, the *orbital period* (*P*) in years, the *time of periastron passage* (*T*) of the companion expressed as a year and decimal fraction thereof, and the *eccentricity* (*e*) of the true orbit. Three additional "geometric" parameters are needed for a complete solution (which requires knowing the true shape and orientation of the orbit in space as opposed to the apparent one we see). These are the *inclination* of the plane of the true orbit (*i*) in degrees to the plane of the sky, and the *position angle of the ascending node* and *longitude of periastron* (both in degrees), designated by the upper and lower case Greek letter omega, respectively.

Obviously, determining the orbital elements of visual binary stars is an involved and complex process. Each of the references cited above under micrometer measurement of doubles offers help on how to go about it. In recent years, a number of software programs for performing these calculations have become available both

Figure 6.10. A video image of Porrima taken by Rainer Anton of Germany with his homemade 10-inch Newtonian reflector shown in **Fig. 5.8**. Long an easy and attractive pair for small telescopes, it's now rapidly closing up on its way to periastron, at which time it will lie beyond the reach of all but the largest amateur instruments. Courtesy Rainer Anton.

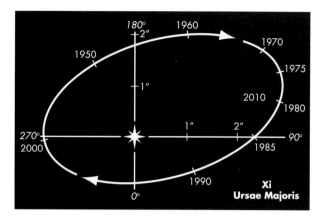

Figure 6.11. Orbit of the celebrated visual binary ξ Ursae Majoris, the first double star to have its orbital path calculated. With a period of 60 years, the companion has now made nearly four circuits of the primary since its discovery by Sir William Herschel in 1780. Courtesy *Sky & Telescope*.

as freeware (or shareware) on the Internet and as commercial packages. They have helped take much of the drudgery and complexity out of this work. Yet most amateurs are content to simply submit their data to the professional for analysis and devote their efforts to making the needed observations at the telescope.

Whether you choose to leave this work to the specialist or attempt to do the orbital calculations yourself, you should take time to plot the observed separation and position angle of pairs against ime to find their apparent orbits. (There exist some very elegant older graphical methods for finding the true orbit, but these are not used much today.) Except at the time of periastron passage for fast-moving pairs like Porrima, you will in most cases observe only a very small portion of the orbit over a period of years. But it is still fun to attempt to plot the path of companion stars about their primaries from these little arcs, and it keeps the observer going back to the telescope again and again to make additional measures to add to the graph.

New Pair Survey

With all the great visual surveys of the sky conducted in the past to discover double stars, it would seem to be a waste of time to attempt to find any new ones today. But the key word here is "past" – these surveys were in most cases made nearly a century or more ago. This, combined with the fact that binary stars are constantly in motion, opens up the possibility that some pairs were too close to resolve back then but have since opened up and become visible (or were perhaps entirely missed in the first place.) This is believed to have been the case in the surprising 1980 discovery that the famous double Albireo is actually a visual triple system. While this was detected using an observatory-class instrument, a number of amateur astronomers in recent years have discovered companions to nearly a hundred stars that were previously thought to be single. Among the telescopes used in these observations were ones as small as a 10-inch reflector, a 9-inch refractor and an 8-inch Schmidt-Cassegrain.

Some astrometric and spectroscopic binaries have had their companions unexpectedly detected visually when at maximum separation in their orbits using large telescopes, and these might make good target candidates. It is also suspected that many doubles lurk undetected among the fainter stars – those of 10th magnitude and below. And the brighter stars have their own share of surprises too, as seen above. So a systematic visual survey for new pairs might indeed be a worthwhile project for those who enjoy the thrill of discovery. A high-quality, large-aperture telescope, excellent sky conditions (preferably *both* seeing and transparency), keen vision, and lots of patience in eliminating known pairs are the primary requirements for such an endeavor.

Founding a Double Star Observers' Society

A final project suggests itself. Again, although technically not really an observational activity itself, it is one which nevertheless has great potential value to observers. Perhaps some zealous lover of these tinted jewels and waltzing couples of the sky will form an international society of double and multiple star observers. Its primary purpose would be to promote and coordinate visual observation of these objects by amateur astronomers, much as the American Association of Variable Star Observers (AAVSO) and the British Astronomical Association (BAA) have done for variable stars. It could help foster communication among double star observers on-line using the Internet, and/or though its own journal devoted to the society's activities. Double star observations by members, as well as news of related work and latest discoveries in this field, could be published on a regular basis through either of these mediums.

A step in this direction was taken a number of years ago by American amateur Ronald Tanguay, whose work was referenced above. He founded the Association of Binary Stars Observers in 1990, primarily for those interested in measuring double stars. While this has since disbanded, he continues to edit *The Double Star Observer*, its bi-monthly journal, which publishes measurements made by both amateur and professional observers from around the world. As mentioned above, his *Double Star Observer's Handbook* also contains a wealth of valuable information on all aspects of these objects. More information about both the journal and the book can be found at his web site http://home.cshore.com/rfroyce/dso/ or by mail at 306 Reynolds Drive, Saugus, MA 01906 USA.

While also not a double star society as such, the British-based Webb Society (named after the Reverend T.W. Webb, author of the classic work *Celestial Objects for Common Telescopes* which first appeared in 1859) publishes double star measurements and related observations in its quarterly journal, *The Deep-Sky Observer*. This organization has also issued a number of excellent deep-sky observing guidebooks, one of which is *Double Stars* by Robert Argyle. The Society can be contacted on the Internet at www.webbsociety.freeserve.co.uk.

The need for a double star observers' society clearly exists. Its birth could well breathe new life into an important but long-neglected field of fundamental observational astronomy – one that possesses a proud heritage from the past and holds great promise for the future.

Chapter 7

Double and Multiple Star Observing Lists

In this chapter we present the first of two double and multiple star observing lists. It is a compilation of a hundred of the finest showpieces for viewing with telescopes from 2- to 14-inches in aperture. Nearly all of them can be seen in the smallest of glasses, and some even in binoculars. Arranged alphabetically by constellation (which makes it more convenient to pick out objects for a given night's observations than one ordered by coordinates), it features brief descriptions of each entry. The second roster of 400 pairs is intended for those who want to see more of these starry jewels – and/or pursue some of the projects suggested in the previous chapter, such as revising Dawes' Limit or making micrometer measures of binaries. This working table *is* arranged by Right Ascension rather than by constellation and appears in Appendix 3. Together, these lists offer 500 selected objects for exploration and study.

Primary data sources for both lists were *Sky Catalogue 2000.0* and the *Washington Double Star Catalog*. Right Ascension (RA) in hours and minutes, and Declination (Dec) in degrees and minutes, are for the current standard Epoch 2000.0. Constellation (Con) abbreviations are the official three-letter designations adopted by the International Astronomical Union (see the constellation listing in Appendix 1.) Other table headings are the apparent visual magnitudes (Mags) of the components, their approximate current angular separation (Sep) in arc-seconds, and their spectral types (Spec) on either the standard MKK (Morgan-Keenan-Kellman) system or the HD (Henry Draper) system, if available. (For more information on spectral classes see Michael Inglis's excellent *Observer's Guide to Stellar Evolution*, Springer-Verlag.) Position angles are not given for a variety of reasons (among them the confusion resulting from the common use of star diagonals with refracting and compound telescopes, producing "inside-out" mirror-images of the sky). Those observers desiring the latest available position angles, as well as measures of component separations, should consult the U.S. Naval Observatory's *Washington Double Star Catalog* on-line at http://ad.usno.navy.mil/wds/

Approximate distance in light-years (LY) is also given in many cases. Unless an orbital period is indicated, or a pair is noted as being "optical" (meaning it consists of two unrelated stars that happen to lie along the same line of sight), the objects are common proper motion (or CPM) systems – those drifting through space together and, therefore, gravitationally bound. In most (if not all) cases such pairs are actually very slowly orbiting each other, but in periods measured in thousands of years. Finally, both lists extend down to –45 degrees Declination, covering that three-fourths of the entire heavens visible from mid-northern latitudes.

Double and Multiple Star Observing Lists

Table 7.1. One Hundred Showpiece Double and Multiple Stars

Object/Con	RA	Dec	Mags	Sep	Spec	Name/Description
γ And	02h 04m	+42° 20'	2.3, 5.5, 6.3	10", 0.4"	K3II, B9V, A0V	Almach. Brilliant topaz-orange and aquamarine pair. Superb contrast! B-C tight 61-yr. binary. 300LY
56 And	01h 58m	+37° 15'	5.7, 5.9	190"	K0III, M0III	Golden wide pair near SW edge of big open cluster NGC 752. 360LY
ζ Aqr	22h 29m	−00° 01'	4.4, 4.5	2"	F1IV, F5IV	Matched, bright, off-white close duo. Famous 850-yr. binary. 76LY
94 Aqr	23h 19m	−13° 28'	5.3, 7.3	13"	G5IV, K2V	Pale rose or reddish and light emerald green. Lovely but neglected object.
15 Aql	19h 05m	−04° 02'	5.5, 7.2	38"	K0, K1III	Easy, wide duo. Yellowish-orange and ruddy purple or lilac. Optical.
57 Aql	19h 55m	−08° 14'	5.8, 6.5	36"	B7V, B8V	Another roomy, easy pair. Both stars bluish-white with hint of other hues.
γ Ari	01h 54m	+19° 18'	4.8, 4.8	8"	B9V, A1	Stunning, bright perfectly matched blue-white pair. Superb! 160LY
λ Ari	01h 58m	+23° 36'	4.9, 7.7	37"	F0V, G0	Wide color and mag. contrast double.
θ Aur	06h 00m	+37° 13'	2.6, 7.1	4"	A0II, G2V	Radiant, tight mag. contrast pair for steady nights; lilac and yellow. 110LY
ε Boo	14h 45m	+27° 04'	2.5, 4.9	2.8"	K0II, A2V	Izar. Bright, tight pair, superb pale-orange and sea-green. Struve's "the most beautiful one." 160LY
ξ Boo	14h 51m	+19° 06'	4.7, 7.0	6"	G8V, K4V	Striking yellow and reddish-orange or purple. 150-yr. binary. 22LY
μ Boo	15h 24m	+37° 23'	4.3, 7.0, 7.6	108", 2"	F0V, K0, K0	Neat triple system! B-C is 260-yr. binary. Yellowish and oranges. 95LY
κ Boo	14h 14m	+51° 47'	4.6, 6.6	13"	A8IV, F1V	Pretty double – tints real but elusive.

Table 7.1. One Hundred Showpiece Double and Multiple Stars (continued)

Object/Con	RA	Dec	Mags	Sep	Spec	Name/Description
π Boo	14h 41m	+16° 25'	4.9, 5.8	6"	B9, A6V	Closer version of κ Boo.
ζ Boo	14h 41m	+13° 44'	4.5, 4.6	0.7"	A3IV, A2V	Matched white, ultra-close 125-yr. binary. Elongated egg in small glass.
32 Cam	12h 49m	+83° 25'	5.3, 5.8	22"	A1III, A0V	Nice matched off-white pair. 495LY
ζ Cnc	08h 12m	+17° 39'	5.6, 6.0, 6.2	0.9", 6"	F8V, F9V, G5V	Beautiful close matched trio. 60- and 1150-yr. periods. All yellow. 70LY
ι-1 Cnc	08h 47m	+28° 46'	4.2, 6.6	30"	G8III, A3V	Albireo (β Cyg) of spring… Striking orange and blue pair! 165LY
α CVn	12h 56m	+38° 19'	2.9, 5.5	20"	B9-A0p, F0V	Cor Caroli. Magnificent blue-white double – one of the finest. 130LY
α CMa	06h 45m	−16° 43'	−1.46, 8.5	6"	A1V, WDA	Sirius. Blazing blue-white sapphire with famed white-dwarf companion. A 50-yr. binary now widening. 9LY
h 3945 CMa	07h 17m	−23° 19'	4.8, 6.8	27"	K3I, F0	Albireo (β Cyg) of winter. Splendid orange and blue duet!
α–1/2 Cap	20h 18m	−12° 33'	3.6, 4.2	378"	G9III, G3I	Naked-eye orange pair with 10.6- and 9.6-mag. companions at 7" and 45" forming a telescopic double-double. Optical – 110LY and 700LY!
η Cas	00h 49m	+57° 49'	3.4, 7.5	13"	G0V, M0V	Beautiful yellow and ruddy-purple or garnet combo. 480-yr. binary – just 19LY away.
ι Cas	02h 29m	+67° 24'	4.6, 6.9, 8.4	2.5", 7"	A5, F5, G5	Elegant but tough triple system. Hues yellow, lilac and blue. Close pair 840-yr. binary. 160LY
σ Cas	23h 59m	+55° 45'	5.0, 7.1	3"	B1V, B3V	Tight pair, intense bluish and greenish tints. Quite distant – 1400LY

Double and Multiple Star Observing Lists

Table 7.1. One Hundred Showpiece Double and Multiple Stars (continued)

Object/Con	RA	Dec	Mags	Sep	Spec	Name/Description
β Cep	21h 29m	+70° 34'	3.2, 7.9	13"	B2III, A2V	Neat unequal pair – greenish-white and blue or purple. Exquisite! 980LY
δ Cep	22h 29m	+58° 25'	3.5–4.4, 6.3	41"	F5I-GII, B7	Striking pale orange and blue gems. Primary prototype of Cepheid variables – period 5.4 days. 1300LY
ξ Cep	22h 04m	+64° 38'	4.4, 6.5	8"	A3, F7	Neat bright pair, subtle colors. 80LY
γ Cet	02h 43m	–03° 14'	3.5, 6.2	3"	A3V, F3	Close, bright; yellow and ashen. 63LY
24 Com	12h 35m	+18° 23'	5.2, 6.7	20"	K2III, A9V	Vivid orange and blue-green duo – an intensely hued lovely jewel! 300LY
ζ CrB	15h 39m	+36° 38'	5.1, 6.0	6"	B7V, B7V	Pretty pair of bluish and greenish suns.
σ CrB	16h 15m	+33° 52'	5.6, 6.6	7"	G0V, G1V	Like ζ but stars are yellowish. Binary with 1000-yr. period.
δ Crv	12h 30m	–16° 31'	3.0, 8.4	24"	B9V, K2V	Nice mag. and color contrast combo. Yellow and violet or lilac. 125LY
β Cyg	19h 31m	+27° 58'	3.1, 5.1	34"	K3II+B9V, B8V	Albireo. One of grandest sights in the heavens! Magnificent topaz and sapphire-blue pair in radiant Milky Way setting. Finest double. 380LY
o-1 Cyg	20h 14m	+46° 44'	3.8, 7.7, 4.8	107", 338"	K2II, B9, A5III	Wide neat trio – orange, blue, white, in another sparkling setting. 200LY
δ Cyg	19h 45m	+45° 08'	2.9, 6.3	2.5"	B9III, F1V	Bright, close unequal pair – tough! Greenish-white and ashen. Binary with 800-yr. period. 270LY
16 Cyg	19h 42m	+50° 32'	6.0, 6.1	39"	G2V, G5V	Lovely wide, matched golden duo in wide field with Blinking Planetary (NGC 6826).
61 Cyg	21h 07m	+38° 45'	5.2, 6.0	30"	K5V, K7V	Beautiful easy orange pair. Famous as first star to have distance (11LY) measured. Slow 650-yr. binary.

Table 7.1. One Hundred Showpiece Double and Multiple Stars (continued)

Object/Con	RA	Dec	Mags	Sep	Spec	Name/Description
γ Del	20h 47m	+16° 07'	4.5, 5.5	10"	K1IV, A2I	Stunning golden-yellow and greenish-blue combo – splendid sight! 100LY
μ Dra	17h 05m	+54° 28'	5.7, 5.7	2"	F7V, F7V	Cozy, yellowish-white identical-twin binary with 480-yr. period. 82LY
ν Dra	17h 32m	+55° 11'	4.9, 4.9	62"	A5, A5	Another pair of perfectly-matched suns, but brighter and much wider than μ. Both white – superb! 120LY
ψ Dra	17h 42m	+72° 09'	4.9, 6.1	30"	F5IV, G0V	Pretty yellow and lilac pair – easy.
17/16 Dra	16h 36m	+52° 55'	5.4, 6.4, 5.5	3", 90"	B9V, A1V, B9V	Nice triple system like μ Boo but primary has comp. All white.
41/40 Dra	18h 00m	+80° 00'	5.7, 6.1	19"	F7, F7	Pale-yellow pair with 7.5-mag. near.
θ Eri	02h 58m	–40° 18'	3.4, 4.5	8"	A4III, A1V	Radiant white, far-south gems! 93LY
32 Eri	03h 54m	–02° 57'	4.8, 6.1	7"	G8III, A2V	Lovely topaz-yellow and sea-green in superb contrast – a beauty! 300LY
o-2 Eri	04h 15m	–07° 39'	4.4, 9.5, 11.2	83", 8"	K1V, DA, M4	Faint pair, an amazing white-dwarf and red-dwarf 248-yr. binary. 16LY
α Gem	07h 35m	+31° 53'	1.9, 2.9, 8.9	4", 72"	A1V, A2V, M0V	Castor. Dazzling blue-white 470-yr. binary – magnificent sight! Orange comp. is eclipsing pair YY Gem, ranging from mag. 8.9 to 9.6 over 20 hours. All one system. 52LY
δ Gem	07h 20m	+21° 59'	3.5, 8.2	6"	F0IV, K3V	Yellow and reddish-purple duo. A 1200-yr. binary. 53LY
α Her	17h 15m	+14° 23'	3.1–3.9, 5.4	5"	M5II, G5III	Rasalgethi. Bright, intensely tinted orange and blue-green pair – superb! Primary huge pulsating semiregular variable – a supergiant sun. Orbital period estimated at 3600 yrs. 380LY

Double and Multiple Star Observing Lists

Double and Multiple Star Observing Lists

Table 7.1. One Hundred Showpiece Double and Multiple Stars *(continued)*

Object/Con	RA	Dec	Mags	Sep	Spec	Name/Description
δ Her	17h 15m	+24° 50'	3.1, 8.7	14"	A3IV, G4	Famec, very delicate optical pair. White and violet. 94LY
κ Her	16h 08m	+17° 03'	5.3, 6.5	28"	G8III, K1III	Striking yellow and garnet jewels.
ρ Her	17h 24m	+37° 09'	4.6, 5.6	4"	A0V, B9III	Bright, cozy bluish and greenish pair.
95 Her	18h 02m	+21° 36'	5.0, 5.1	6"	A5III, G8III	Lovely twin suns – amazing "apple-green and cherry-red" tints! 380LY
100 Her	18h 08m	+26° 06'	5.9, 6.0	14"	A3V, A3V	Another matched pair but wider than 95 Her with pale off-white hues. Little known.
ε Hya	08h 47m	−06° 25'	3.3, 6.8	3"	F0V+KIII, F5	Tight 890-yr. binary. Primary is also visual binary in observatory scopes, having period of just 15 yrs. 150LY
N Hya	11h 32m	−29° 16'	5.8, 5.9	9"	F5, F5	Yellowish-white twins. Very nice!
8 Lac	22h 36m	+39° 38'	5.7, 6.5, 10.5, 9.3	22", 49", 82"	B1V, B2V	Blue-white duo – fainter companions form delicate quadruple. 1900LY
α Leo	10h 08m	+11° 58'	1.4, 7.7	177"	B7V, K1V	Regulus. Wide mag. contrast pair with blue-white primary and indigo comp. 78LY
γ Leo	10h 20m	+19° 51'	2.2, 3.5	4"	K0III, G7III	Algieba. Magnificent radiant golden suns forming 620-yr. binary. One of finest in the heavens. 170LY
54 Leo	10h 56m	+24° 45'	4.5, 6.3	6"	A1V, A2V	Lovely, little-known bluish-white and greenish-white pair. 150LY
γ Lep	05h 44m	−22° 27'	3.7, 6.3	96"	F6V, K2V	Pretty, wide pale-yellow and garnet combo awash in color. 29 LY
12 Lyn	06h 46m	+59° 27'	5.4, 6.0, 7.3	1.7", 9"	A3V	Fascinating tight trio, all white. A-B 700-yr. binary. 140LY
38 Lyn	09h 19m	+36° 48'	3.9, 6.6	3"	A1V, A4V	Bright close pair with subtle tints.

Table 7.1. One Hundred Showpiece Double and Multiple Stars *(continued)*

Object/Con	RA	Dec	Mags	Sep	Spec	Name/Description
α Lyr	18ʰ 37ᵐ	+38° 47'	0.0, 9.5, 9.5	63", 118"	A0V	Vega. Dazzling pale-sapphire gem with faint comps. Beautifull 26LY
β Lyr	18ʰ 50ᵐ	+33° 22'	3.3–4.3, 8.6 8.6, 9.9, 9.9	46", 67", 86"	B7V+A8	Famous 12.9-day eclipsing binary set within starry triangle. 860LY
ε-1/2 Lyr	18ʰ 44ᵐ	+39° 40'	5.0, 6.1 5.2, 5.5	2.6" 2.3"	B2V, F1V A8V, F0V	Famed "Double-Double" multiple system. Pairs 207" apart. Closer one 600-yr. and wider one 1200-yr. binary slowly orbiting each other. Stars all white. Wondrous sight! 200LY
ζ Lyr	18ʰ 45ᵐ	+37° 36'	4.3, 5.9	44"	A0, F0IV	Easy topaz and pale-green double.
δ Lyr	18ʰ 54ᵐ	+36° 58'	4.5, 5.6	630"	M4II, B2V	Ultra-wide but lovely reddish-orange and blue pair involved in sparse open cluster Stephenson-1. Both 800LY
β Mon	06ʰ 29ᵐ	−07° 02'	4.7, 5.4, 5.6	3", 10"	B3V, B3, B3	Herschel's Wonder Star. Superb trio, all bluish-white, in slender triangle. Fascinating spectacle! 700LY
ε Mon	06ʰ 24ᵐ	+04° 36'	4.5, 6.5	13"	A5IV, F5V	Pretty gold and blue pair in rich field.
36 Oph	17ʰ 15ᵐ	−26° 36'	5.1, 5.1, 6.7	5", 730"	K0V, K1V, K5	Pretty perfectly matched close pair, a 550-yr. binary, with wide comp. All golden-orange and connected. 18LY
o Oph	17ʰ 18ᵐ	−24° 17'	5.4, 6.9	10"	K0II, F6IV	Lovely orange and clear-blue jewels.
70 Oph	18ʰ 06ᵐ	+02° 30'	4.2, 6.0	4"	K0V, K6	Famous yellow and red binary with 88-yr. period. Superb pair! 17LY
β Ori	05ʰ 14ᵐ	−08° 12'	0.1, 6.8	10"	B8I, B5V	Rigel. Beautiful blue-white super-giant sun with fainter comp. forming splendid mag. contrast pair! 900LY
η Ori	05ʰ 25ᵐ	−02° 24'	3.1–3.4, 4.8	1.5"	B1V+B2, B?	Bright, tight bluish-white duo whose primary is an eclipsing binary with 8-day period. 1400LY
λ Ori	05ʰ 35ᵐ	+09° 56'	3.6, 5.5	4"	O8III, B0V	Neat cozy pair, both bluish-white with hint of violet or purple. 900LY

Double and Multiple Star Observing Lists

Table 7.1. One Hundred Showpiece Double and Multiple Stars (continued)

Object/Con	RA	Dec	Mags	Sep	Spec	Name/Description
δ Ori	$05^h 32^m$	$-00°\ 18'$	1.9-2.1, 6.3	53"	O9II, B2V	Wide mag. contrast pair with 5.7-day eclipsing primary. Greenish-white and pale-blue or violet comp. 1400LY
ζ Ori	$05^h 41^m$	$-01°\ 57'$	1.9, 4.0	2.5"	O9I, B0III	Bright close blue-white duo. 1400LY
σ Ori	$05^h 39^m$	$-02°\ 36'$	4.0, 10.3 7.5, 6.5	11" 13", 43"	O9V B2V, B2V	Amazing colorful multiple star with faint triple Σ761 (8.0, 8.5, 9.0, 68", 8") in field; all one system! Many diverse hues evident. 1200LY
ι Ori	$05^h 35^m$	$-05°\ 55'$	2.8, 6.9	11"	O5, B9	Diamondlike pair with Σ 747 (4.8, 5.7, 36", both B1) in same radiant gem-field – forming a wide double-double system! 1400LY
θ-1 Ori	$05^h 35^m$	$-05°\ 23'$	6.4, 7.9 5.1, 6.7	9" 13", 22"	B0V, B0V O6, B0V	Famed "Trapezium" multiple star embedded in the heart of the Orion Nebula – a magnificent spectacle. Like diamonds against green velvet. Also several fainter companions, forming a small star cluster. 1600LY
η Per	$02^h 51^m$	$+55°\ 54'$	3.8, 8.5	28"	K3I, B?	Nice color and mag. contrast pair with vivid orange and blue hues.
α Psc	$02^h 02^m$	$-02°\ 46'$	4.2, 5.1	1.8"	A0, A3	Alrescha. Tight double with strange, subtle tints. 720-yr. binary. 130LY
ψ-1 Psc	$01^h 06^m$	$+21°\ 28'$	5.6, 5.8	30"	A1V, A0V	Easy matched pair, both blue-white.
ζ Psc	$01^h 14^m$	$+07°\ 35'$	5.6, 6.5	23"	A7IV, F7V	Pale-yellow and pale lilac duo. 140LY
k Pup	$07^h 39^m$	$-26°\ 48'$	4.5, 4.7	10"	B6V, B5IV	Superb bright double sun resembling γ Ari. Both blue-white.
α Sco	$16^h 29^m$	$-26°\ 26'$	0.9-1.8, 5.4	2.5"	M2I, B3V	Antares. Beautiful fiery-red supergiant with emerald-green comp. for steady nights. 900-yr. binary. 520LY
β Sco	$16^h 05^m$	$-19°\ 48'$	2.6, 4.9	14"	B0V, B2V	Graffias. Radiant blue-white combo resembling famed Mizar (ζ UMa). A beautiful sight! 600LY

Table 7.1. One Hundred Showpiece Double and Multiple Stars *(continued)*

Object/Con	RA	Dec	Mags	Sep	Spec	Name/Description
ν Sco	16h 12m	−19° 28′	4.3, 6.8 6.4, 7.8	0.9″ 2.3″	B3V, B9 B8V, B9V	Tight double-double with pairs 41″ apart. Subtle tints. Neat! 440LY
ξ Sco	16h 04m	−11° 22′	4.8, 7.3	8″	F5IV, F5IV	Yellow pair with Σ 1999 (7.4, 8.1, 12″, G8V, K5III) 280″ away forming wide double-double. Primary 46-yr. close binary. 80LY
δ Ser	15h 35m	+10° 32′	4.2, 5.2	4″	F0IV, F0IV	Stunning, neatly paired double with off-white hues. Elegant! 85LY
θ Ser	18h 56m	+04° 12′	4.5, 5.4	22″	A5V, A5V	Wider version of δ Ser. A very pretty, easy pair! 140LY
θ-1/2 Tau	04h 29m	+15° 52′	3.4, 3.8	337″	A7III, G7III	Neat, bright naked-eye and binocular combo in Hyades cluster near topaz Aldebaran. White and yellow. 150LY
ι Tri	02h 12m	+30° 18′	5.3, 6.9	4″	G5III+F5V, F6V	Tight gold and blue-green pair. 200LY
ζ UMa	13h 24m	+54° 56′	2.3, 4.0, 4.0	14″, 709″	A2V, A1, A5V	Famed Mizar with Alcor nearby; trio of radiant blue-white diamonds. One of the finest objects in the heavens. All three are unresolved spectroscopic binaries (like many others here) and one vast sextuple system. 78LY
ξ UMa	11h 18m	+31° 32′	4.3, 4.8	1.8″	G0V, G0V	First binary to have its orbital period determined (60-yrs.) – has made over three circuits since discovery. Twin yellowish suns in contact. 26LY
α UMi	02h 32m	+89° 16′	1.9–2.1, 9.0	18″	F7I, F1	Polaris. Unequal mag. contrast pair having amazing (apparent) "24-hour orbital period" caused by the Earth's rotation. Primary brightest Cepheid in the sky. May be optical. 430LY
γ Vel	08h 10m	−47° 20′	1.8, 4.3	41″	WC8+O8, B1IV	Although over 2 degrees below our Dec cutoff, this dazzling pair is one of most beautiful in the heavens and worth every effort to see! 1000LY

Double and Multiple Star Observing Lists

Double and Multiple Star Observing Lists

Table 7.1. One Hundred Showpiece Double and Multiple Stars (continued)

Object/Con	RA	Dec	Mags	Sep	Spec	Name/Description
γ Vir	12ʰ 42ᵐ	−01° 27′	3.5, 3.5	1.5″	F0V, F0V	Porrima. Famed bright binary with 171-yr. period. Now closing to 0.4″ minimum in 2005, these suns can be seen merging into one yellowish egg with slowly turning long axis. 39LY

Chapter 8

Conclusion

We have now come to the end of our journey of exploration and discovery of visual double and multiple stars. In the first part of this book, we provided an introduction to the subject – one that the author hopes conveyed the many joys and excitement of viewing this class of deep-sky wonders. We also examined the many fascinating types of double stars to be encountered among the stellar population. Finally, we turned our attention to various astronomical and astrophysical considerations involving these plentiful stellar systems.

In the second part of the book, we explored both techniques for observing these objects and the telescopes and accessories needed for doing so, followed by a number of suggested double star observing projects. We concluded by providing an exciting roster of 100 showpiece targets for viewing with typical amateur instruments. And now we bring it all together in our closing remarks aimed at creating a proper perspective of what double star observing is really all about.

Reporting and Sharing Observations

While observing double and multiple stars can be a rewarding activity on a strictly personal basis and can be enjoyed alone, it is in sharing your observations with other people and organizations that its pursuit realizes its full value. As the old English proverb states, "A joy that's shared is a joy made double." At its simplest level, letting family, friends, neighbors, beginning stargazers – and even total strangers passing by – see wonders like the magnificent pair Albireo through your telescope will not only bring delight to them, but also immense satisfaction to yourself. Perhaps it will be *you* who first opens up a totally unexpected and awesome new universe to them!

Sharing observations by reporting them to one of several double star organizations takes you a step beyond this basic (but very important!) level into the area of potentially contributing to our knowledge and understanding of these objects in the grand cosmic scheme of things. There are currently two main groups coordinating and reporting double star observations, one in the U.K. and the other in the United States.

The British-based Webb Society, named after the Reverend T.W. Webb – author of the classic work *Celestial Objects for Common Telescopes* that first appeared in 1859 – publishes double star measurements and related observations in its quarterly journal, *The Deep-Sky Observer*. This organization has also issued a number of excellent observing guidebooks, Volume 1 of which is entitled *Double Stars* by Robert Argyle. The Society can be contacted on the Internet at www.webb-society.freeserve.co.uk.

The Double Star Observer is the title of an excellent bi-monthly journal published and edited by Massachusetts amateur astronomer Ronald Charles Tanguay, whose amazing work measuring double stars with a modified reticle eyepiece micrometer using a 3.5-inch Maksutov telescope was cited in Chapter 6. Although the double star society he founded a over decade ago – known as the Association for Binary Star Observers – has since disbanded, its work is being carried on by the journal itself. Its home page is filled with a vast amount of information on double stars and can be accessed on-line at http://www.home.cshore.com/rfroyce/dso/. Tanguay himself can be contacted by mail at 306 Reynolds Drive, Saugus, MA 01906 USA. He is also author and publisher of the *Double Star Observer's Handbook*, a superb reference for anyone interested in these objects.

Another venue for sharing your observations could soon make its appearance in the field of astronomy. It would be a true international double and multiple star observer's society – one personally founded by *you* should you decide to follow up on the suggestions made in Chapter 6. If you do, the astronomical community would truly be in your debt, and a legion of great double star observers from the past would surely be smiling down upon your efforts.

Figure 8.1. American double star enthusiast Sissy Haas, looking through her modified "department-store" 2.4-inch (60 mm) refractor. Some of her lovely eyepiece drawings of doubles appear in Chapter 1 and Chapter 6 of this book, while many more have graced the pages of *Sky & Telescope* magazine over the years. She typifies the casual (but active!) visual observer of these tinted jewels of the heavens. Courtesy Sissy Haas.

Pleasure versus Serious Observing

Rampant in the field of amateur astronomy today is a belief that you must be doing "serious work" of value to science, preferably with sophisticated (and, therefore, expensive) telescopes and accessories, in order to call yourself an observer. This is truly unfortunate, for it has undoubtedly discouraged many budding stargazers from pursuing astronomy as a pastime.

The root of the word "amateur" is the Latin word "amare" – which means "to love" – or more precisely, from "amator", which means "one who loves." An amateur astronomer is one who loves the stars – loves them for the sheer joy of knowing them. He or she may in time come to love them so much that there will be a desire to contribute something to our knowledge of them. But this rarely happens initially – first comes a period of time (for some, a lifetime) getting to know the sky and enjoying its treasures and wonders as a celestial sightseer.

Certainly one of the best examples of this is the American observer Leslie Peltier, who developed an early love of the stars. He went on to discover or co-discover 12 comets and 6 novae, and he contributed over 100,000 visual magnitude estimates of variable stars to the American Association of Variable Star Observers (AAVSO). Born on a farm, where he made his early observations as a boy and young man, he stayed in his rural community and close to nature the remainder of his life – this, despite

Figure 8.2. American observer Ronald Tanguay, editor of the periodical *Double Star Observer* and author of *The Double Star Observer's Handbook*, looking through his 90 mm Maksutov–Cassegrain telescope. Typifying the serious observer, he has almost single-handedly revived interest in the measurement of visual binary stars by amateurs (particularly in the United States) for more than a decade. Courtesy Ronald Tanguay.

offers to join the staffs of several professional observatories! If there is any reader of this book who has not already done so, the author urges you to obtain and devour Peltier's inspiring autobiography *Starlight Nights* (Sky Publishing Corporation).

The well-known British amateur James Muirden, in his *The Amateur Astronomer's Handbook* (Harper & Row), discusses many areas of observational astronomy in which amateurs can become actively involved in programs of potential value to the field. Yet, he wisely advises his readers to "also never forget that astronomy loses half its meaning for the observer who never lets his telescope range across the remote glories of the sky 'with an uncovered head and humble heart'." He goes on to lament that "The study of the heavens from a purely aesthetic point of view is scorned in this technological age." How very sad!

Aesthetic and Philosophical Considerations

For many it is the aesthetic and philosophical (and, for some, spiritual) aspects of astronomy that constitute its greatest value to the individual. For the double star observer in particular, there is perhaps no field of observational astronomy so filled with inspiring vistas of radiant beauty, infinite diversity and heavenly-hued pageantry than that in the realm of double and multiple stars, nor one in which the amateur can so readily contribute valued fundamental data and join ranks with the professional astronomer.

We conclude by offering a selection of thoughts from great observers of the past and present, both amateur and professional, for contemplation. As you let these words penetrate your mind and heart, you will come to find that approach to observing the heavens which is ideally suited for you in your on-going cosmic adventure as "one who loves" the stars!

> "But let's forget the astrophysics and simply enjoy the spectacle." – Walter Scott Houston

> "I became an astronomer not to learn the facts about the sky but to feel its majesty." – David Levy

> "I am because I observe." – Thaddeus Banachiewicz

> "Even if there were no practical application of visual observing, it would always be a sublime way to spend a starry night." – Lee Cain

> "To me, astronomy means learning about the universe by looking at it." – Daniel Weedman

> "Astronomy is a typically monastic activity; it provides food for meditation and strengthens spirituality." – Paul Couteau

> "Astronomy has an almost mystical appeal…we should do astronomy because it is beautiful and because it is fun." – John Bahcall

> "The great object of all knowledge is to enlarge and purify the soul, to fill the mind with noble contemplations, and to furnish a refined pleasure." – Edward Everett

"The true value of a telescope is how many people have viewed the heavens through it." – John Dobson

"The appeal of stargazing is both intellectual and aesthetic; it combines the thrill of exploration and discovery, the fun of sight-seeing, and the sheer joy of first-hand acquaintance with incredibly wonderful and beautiful things." – Robert Burnham, Jr.

"Whatever happened to what amateur astronomers really care about – simply enjoying the beauty of the night sky?" – Mark Hladik

"I would rather freeze and fight off mosquitoes than play astronomy on a computer." – Ben Funk

"The sky belongs to all of us. It is glorious and it is free." – Deborah Byrd

"To me, telescope viewing is primarily an aesthetic experience – a private journey in space and time." – Terence Dickinson

"Observing all seems so natural, so real, so obvious. How could it possibly be any other way?" – Jerry Spevak

"As soon as I see a still, dark night developing, my heart starts pounding and I start thinking 'Wow! Another night to get out and search the universe.' The views are so incredibly fantastic!" – Jack Newton

"Take good care of it [your telescope] and it will never cease to offer you many hours of keen enjoyment, and a source of pleasure in the contemplation of the beauties of the firmament that will enrich and ennoble your life." – William Tyler Olcott

"A telescope is a machine that can change your life." – Richard Berry

"The amateur astronomer has access at all times to the original objects of his study; the masterworks of the heavens belong to him as much as to the great observatories of the world. And there is no privilege like that of being allowed to stand in the presence of the original." – Robert Burnham, Jr.

"But it is to be hoped that some zealous lover of this great display of the glory of the Creator will carry out the author's idea, and study the whole visible heavens from what might be termed a picturesque point of view." – T.W. Webb

"But aren't silent worship and contemplation the very essence of stargazing?" – David Levy

"Adrift in a cosmos whose shores he cannot even imagine, man spends his energies in fighting with his fellow man over issues which a single look through this telescope would show to be utterly inconsequential." – Palomar 200-inch Hale telescope dedication

"How can a person ever forget the scene, the glory of a thousand stars in a thousand hues…" – Walter Scott Houston

"Seeing through a telescope is fifty percent vision and fifty percent imagination." – Chet Raymo

"A night under the stars … rewards the bug bites, the cloudy nights, the next-day fuzzies, and the thousand other frustrations with priceless moments of sublime beauty." – Richard Berry

"All galaxies [and other celestial wonders including double stars!] deserve to be stared at for a full fifteen minutes." – Michael Covington

"It is not accident that wherever we point the telescope we see beauty." – R.M. Jones

"You have to really study the image you see in the eyepiece to get all the information coming to you. Taking a peek and looking for the next object is like reading just a few words in a great novel." – George Atamian

"What we need is a big telescope in every village and hamlet and some bloke there with that fire in his eyes who can show something of the glory the worlds sails in." – Graham Loftus

"Someone in every town seems to me owes it to the town to keep one [a telescope!]." – Robert Frost

"The pleasures of amateur astronomy are deeply personal. The feeling of being alone in the universe on a starlit night, cruising on wings of polished glass, flitting in seconds from a point millions of miles away to one billions of lightyears distant … is euphoric." – Tom Lorenzin

"I am a professional astronomer who deeply loves his subject, is continually in awe of the beauty of nature [and] like every astronomer I've ever met, I am evangelistic about my subject." – Frank Bash

"To turn from this increasingly artificial and strangely alien world is to escape from *unreality*. To return to the timeless world of the mountains, the sea, the forest, and the stars is to return to sanity and truth." – Robert Burnham, Jr.

Appendix 1

Constellation Names and Abbreviations

The following table gives the standard International Astronomical Union (IAU) three-letter abbreviations for the 88 officially recognized constellations, together with both their full names and genitive (possessive) cases, and order of size in terms of number of square degrees. Those in **bold** type are represented in the double star lists in Chapter 7 and Appendix 3.

Table A1. Constellation Names and Abbreviations

Abbrev.	Name	Genitive	Size
And	**Andromeda**	**Andromedae**	**19**
Ant	**Antlia**	**Antliae**	**62**
Aps	Apus	Apodis	67
Aqr	**Aquarius**	**Aquarii**	**10**
Aql	**Aquila**	**Aquilae**	**22**
Ara	Ara	Arae	63
Ari	**Aries**	**Arietis**	**39**
Aur	**Auriga**	**Aurigae**	**21**
Boo	**Bootes**	**Bootis**	**13**
Cae	**Caelum**	**Caeli**	**81**
Cam	**Camelopardalis**	**Camelopardalis**	**18**
Cnc	**Cancer**	**Cancri**	**31**
CVn	**Canes Venatici**	**Canum Venaticorum**	**38**
CMa	**Canis Major**	**Canis Majoris**	**43**
CMi	**Canis Minor**	**Canis Minoris**	**71**
Cap	**Capricornus**	**Capricorni**	**40**
Car	Carina	Carinae	34
Cas	**Cassiopeia**	**Cassiopeiae**	**25**
Cen	**Centaurus**	**Centauri**	**9**
Cep	**Cepheus**	**Cephei**	**27**
Cet	**Cetus**	**Ceti**	**4**
Cha	Chamaeleon	Chamaeleontis	79
Cir	Circinus	Circini	85
Col	**Columba**	**Columbae**	**54**
Com	**Coma Berenices**	**Comae Berenices**	**42**
CrA	**Corona Australis**	**Coronae Australis**	**80**
CrB	**Corona Borealis**	**Coronae Borealis**	**73**
Crv	**Corvus**	**Corvi**	**70**
Crt	Crater	Crateris	53
Cru	Crux	Crucis	88

Table A1. Constellation Names and Abbreviations *(continued)*

Abbrev.	Name	Genitive	Size
Cyg	**Cygnus**	**Cygni**	16
Del	**Delphinus**	**Delphini**	69
Dor	Dorado	Doradus	7
Dra	**Draco**	**Draconis**	8
Equ	**Equuleus**	**Equulei**	87
Eri	**Eridanus**	**Eridani**	6
For	**Fornax**	**Fornacis**	41
Gem	**Gemini**	**Geminorum**	30
Gru	**Grus**	**Gruis**	45
Her	**Hercules**	**Herculis**	5
Hor	Horologium	Horologii	58
Hya	**Hydra**	**Hydrae**	1
Hyi	Hydrus	Hydri	61
Ind	Indus	Indi	49
Lac	**Lacerta**	**Lacertae**	68
Leo	**Leo**	**Leonis**	12
LMi	Leo Minor	Leonis Minoris	64
Lep	**Lepus**	**Leporis**	51
Lib	**Libra**	**Librae**	29
Lup	**Lupus**	**Lupi**	46
Lyn	**Lynx**	**Lyncis**	28
Lyr	**Lyra**	**Lyrae**	52
Men	Mensa	Mensae	75
Mic	Microscopium	Microscopii	66
Mon	**Monoceros**	**Monocerotis**	35
Mus	Musca	Muscae	77
Nor	Norma	Normae	74
Oct	Octans	Octantis	50
Oph	**Ophiuchus**	**Ophiuchi**	11
Ori	**Orion**	**Orionis**	26
Pav	Pavo	Pavonis	44
Peg	**Pegasus**	**Pegasi**	7
Per	**Perseus**	**Persei**	24
Phe	**Phoenix**	**Phoenicis**	37
Pic	Pictor	Pictoris	59
Psc	**Pisces**	**Piscium**	14
PsA	**Piscis Austrinus**	**Piscis Austrini**	60
Pup	**Puppis**	**Puppis**	20
Pyx	**Pyxis**	**Pyxidis**	65
Ret	Reticulum	Reticuli	82
Sge	**Sagitta**	**Sagittae**	86
Sgr	**Sagittarius**	**Sagittarii**	15
Sco	**Scorpius**	**Scorpii**	33
Scl	**Sculptor**	**Sculptoris**	36
Sct	**Scutum**	**Scuti**	84
Ser	**Serpens**	**Serpentis**	23
Sex	**Sextans**	**Sextantis**	47
Tau	**Taurus**	**Tauri**	17
Tel	Telescopium	Telescopii	57
Tri	**Triangulum**	**Trianguli**	78
TrA	Triangulum Australe	Trianguli Australis	83

Table A1. Constellation Names and Abbreviations *(continued)*

Abbrev.	Name	Genitive	Size
Tuc	Tucana	Tucanae	48
UMa	**Ursa Major**	**Ursae Majoris**	**3**
UMi	**Ursa Minor**	**Ursae Minoris**	**56**
Vel	**Vela**	**Velorum**	**32**
Vir	**Virgo**	**Virginis**	**2**
Vol	Volans	Volantis	76
Vul	**Vulpecula**	**Vulpeculae**	**55**

Appendix 2

Double Star Designations

Presented here is an alphabetical listing of all known double and multiple star designations, dating from the earliest reported discoveries in the mid-1600s up to the present time. By long-standing tradition, a double star is "named" for the person who either discovers it or first makes measures of it, the name itself usually being abbreviated or denoted by a symbol followed by a running serial number from that observer's list, catalog or observatory where he worked.

The great *Index Catalogue of Visual Double Stars* (or *IDS*), originally compiled at the Lick Observatory in 1963, contained essentially all previous discovery catalogues. It eventually became the basis for the *Washington Double Star Catalog* (or *WDS*), which is maintained at the United States Naval Observatory and is now the world's standard such reference. (Chapter 5 contains more about it, including how to access it on-line.) The one-, two- and three-letter codes below are those used in these two works. **Bold** entries are ones containing many of those pairs of particular interest to amateur astronomers and within range of their instruments.

Table A2. Codes for Star Designations

Designation	IDS/WDS Code	Discoverer or Observatory
A	**A**	**R.G. Aitken**
Abt	ABT	Giorgio Abetti
AbH	ABH	H.A. Abt
AC	**AC**	**Alvan Clark**
AG	AG	*Astronomische Gesellschaft Katalog*
AGC	AGC	Alvan G. Clark
AlbO	ALB	Albany Observatory
Ald	ALD	H.L. Alden
AlgO	ALG	Algiers Observatory
Ali	ALI	A. Ali
All	ALL	R.M. Aller
Anj	ANJ	J.A. Anderson
Ara	ARA	S. Aravamudan
Ard	ARD	S. Arend
Arg	ARG	F.W.A. Argelander
Arn	ARN	Dave Arnold
Ary	ARY	Robert Argyle
B	B	W.H. van den Bos
Bal	BAL	R. Baillaud

Table A2. Codes for Star Designations *(continued)*

Designation	IDS/WDS Code	Discoverer or Observatory
Bar	BAR	E.E. Barnard
Baz	BAZ	Paul Baize
Bond	BDW	W.C. Bond
Bem	BEM	A. Bemporad
Bes	BES	F.W. Bessel
Bgh	BGH	S. van den Berg
Bha	BHA	T.P. Bhaskavan
Big	BIG	G. Bigourdan
Bird	BRD	F. Bird
Bll	BLL	R.S. Ball
Blo	BLO	M. Bloch
Boo	BOO	S. Boothroyd
Bot	BOT	G. von Bottger
Bra	BRA	M. Brashear
Brt	BRT	S.G. Barton
BrsO	BSO	Brisbane Observatory
Btz	BTZ	E. Bernewitz
β	**BU**	**S.W. Burnham**
βpm	BUP	S.W. Burnham's proper motion catalogue
Che	CHE	P.S. Chevalier
Chr	CHR	Center for High Resolution Astronomy
Cog	COG	W.A. Cogshall
Com	COM	G.C. Comstock
CorO	COO	Cordoba Observatory
Cou	COU	Paul Couteau
CPD	CPD	Cape Photographic Durchmusterung
CapO	CPO	Cape Observatory
Cru	CRU	L. Cruls
Ctt	CTT	Jean-Francois Courtot
CamU	CUA	Cambridge University
Dem	D	Ercole Dembowski
Da	**DA**	**W.R. Dawes**
Dal	DAL	J.A. Daley
Dan	DAN	Andre Danjon
δ	DAW	B.H. Dawson
DrbO	DEO	Dearborn Observatory
Deu	DEU	A.J. Deutsch
Dick	DIC	J. Dick
Dju	DJU	P. Djurkovic
Dob	DOB	W.A Doberck
Doc	DOC	D.J. Docobo
Dom	DOM	Jean Dommanget
Don	DON	H.F. Donner
Doo	DOO	Eric Doolittle
DorO	DOR	Dorpat Observatory
Δ	DUN	J. Dunlop
Dur	DUR	M.V. Duruy
Dyer	DYR	E.R. Dyer, Jr.

Table A2. Codes for Star Designations *(continued)*

Designation	IDS/WDS Code	Discoverer or Observatory
Edd	EDD	Arthur Stanley Eddington
Edg	EDG	D.W. Edgecomb
Egg	EGG	O.J. Eggen
Elt	ELT	G.A. Elliott
Enc	ENC	J.F. Encke
Eng	ENG	R. Engelmann
Es	**ES**	**T.E.H. Espin**
Fab	FAB	C. Fabricius
φ	FIN	W.S. Finsen
Fla	FLA	Camille Flammarion
Fle	FLE	J.O. Fleckenstein
For	FOR	L. Forgeron
Fox	FOX	Philip Fox
Fra	FRA	R. Frangetto
Frh	FRH	R. Furuhjelm
Frk	FRK	W.S. Franks
Frz	FRZ	J. Franz
Fur	FUR	H. Furner
Gallo	GAL	J. Gallo
GAn	GAN	G. Anderson
Gat	GAT	G. Gatewood
Gee	GEE	W.T Geertsen
Gic	GIC	Henry Giclas
Gir	GIR	P.M. Girard
Gale	GLE	W.F. Gale
Gli	GLI	J.M. Gilliss
Glp	GLP	S. de Glasenapp
Gol	GOL	H. Goldschmidt
Grb	GRB	Steven Groombridge
GrnO	GRO	Greenwich Observatory
Gsh	GSH	J. Glaisher
Gtb	GTB	K. Gottlieb
Gui	GUI	J. Guillaume
Gyl	GYL	A.N. Goyal
Gsh	GSH	J. Glaisher
H	**H**	**William Herschel (1782-1784 catalogues)**
HI	**H**	**"-difficult**
HII	**H**	**"-close but measurable**
HIII	**H**	**"-5 to 15 arc seconds separation**
HIV	**H**	**"-15 to 30 arc seconds separation**
HV	**H**	**"-30 to 60 arc seconds separation**
HVI	**H**	**"-60 to 120 arc seconds separation**
HN	**H**	**William Herschel (1821 catalogue)**
Hcw	HCW	H.C. Wilson
HvdO	HDO	Harvard Observatory
HDS	HDS	Hipparcos Double Star
HvdC	HDZ	Harvard Observatory zone catalogues
hei	HEI	Wulff Heintz

Table A2. Codes for Star Designations *(continued)*

Designation	IDS/WDS Code	Discoverer or Observatory
Hill	HIL	L. Hill
HipC	HIP	Hipparcos Catalogue 1997
h	**HJ**	**John Herschel**
Hall	HL	Asaph Hall
Hld	HLD	E.S. Holden
Hlm	HLM	E. Holmes
Hln	HLN	Frank Holden
Ho	HO	G.W. Hough
Hrg	HRG	L. Hargrave
Hrl	HRL	G. Harlan
Hrs	HRS	D.L. Harris
Hrt	HRT	W.I. Hartkopf
Htg	HTG	C.S. Hastings
Hu	HU	W.J. Hussey
Howe	HWE	H.A. Howe
Hzg	HZG	E. Hertzsprung
Hooke	–	Robert Hooke
Huygens	–	Christiaan Huygens
Hynek	–	J. Allen Hynek
I	I	R.T.A. Innes
J	J	Robert Jonckheere
Jc	JC	W.S. Jacob
Jck	JCK	John Jackson
Jef	JEF	H.M. Jeffers
Joy	JOY	Alfred Joy
Jsp	JSP	M.K. Jessup
Kam	KAM	Peter van de Kamp
Klk	KLK	P.G. Kulikovsky
Knt	KNT	G. Knott
Kop	KOP	Z. Kopal
Kr	KR	A. Kruger
Kron	KRO	G. Kron
Kru	KRU	E.C. Kruger
Ku	KU	F. Kustner
Kui	KUI	Gerard Kuiper
L	L	Thomas Lewis
Lac	LAC	G.B. Lacchini
Lal	LAL	F. de Lalande
Lam	LAM	J. von Lamont
Lar	LAR	J. Larink
Lau	LAU	H.E. Lau
Law	LAW	G.K. Lawton
Lbz	LBZ	P. Labitzke
Lcl	LCL	N.L. de Lacaille
LDS	LDS	W.J. Luyten (1st proper motion catalogue)
Lee	LEE	O.J. Lee
Leo	LEO	Frederick Leonard
Lem	LEM	Lembang Observatory

Table A2. Codes for Star Designations *(continued)*

Designation	IDS/WDS Code	Discoverer or Observatory
Ling	LIN	J.F. Ling
Lip	LIP	Sarah Lee Lippincott
LicO	LO	Lick Observatory
Lob	LOB	D.C. Lobao
LplO	LPO	La Plata Observatory
LPM	LPM	W.J. Luyten (2nd proper motion catalogue)
Lsl	LSL	William Lassell
Lyot	LT	B. Lyot
Luy	LUY	W.J. Luyten
Lv	LV	F.P. Leavenworth
Maa	MAA	A. van Maanen
Ma	MA	J.H. Madler
MadO	MDO	Madras Observatory
Mau	MAU	E.W. Maunder
Mca	MCA	Harold McAlister
Mic	MIC	A.A. Michelson
Mil	MIL	J.A. Miller
Mkt	MKT	Mark III Interferometer
Mla	MLA	Dean McLaughlin
Mlb	MLB	W. Milburn
Mlf	MLF	Frank Muller
Mll	MLL	S.A. Mitchell
MlbO	MLO	Melbourne Observatory
Mlr	MLR	Paul Muller
Mrz	MRZ	William Markowitz
Msn	MSN	Brian Mason
Mtl	MTL	O.M. Mitchell
NeS	NES	Simon Newcomb
Nic	NIC	Seth Nickleson
NwzO	NZO	New Zealand Observatory
Ol	OL	C.P. Olivier
Opik	OPI	E.J. Opik
Ost	OST	P.T. Oosterhoff
Par	PAR	J.A. Parkhurst
Per	PER	J. Perrotin
Phl	PHL	T.E. Phillips
Pic	PIC	W.H. Pickering
Pit	PIT	Colin Pither
PulO	PKO	Pulkovo Observatory
Plq	PLQ	E. Paloque
Pnk	PNK	Dale Penkala
Pol	POL	J.A. Pollock
Pop	POP	G.M. Popovic
Pou	POU	M.A. Pourteau
Prn	PRN	C.D. Perrine
PerO	PRO	Perth Observatory
Pry	PRY	J.J.M. Perry
Prz	PRZ	E. Przbyllok

Table A2. Codes for Star Designations *(continued)*

Designation	IDS/WDS Code	Discoverer or Observatory
Ptt	PTT	Edison Pettit
Pz	PZ	Giuseppi Piazzi
R	R	H.C. Russell
Rei	REI	Karl Reinmuth
Rhd	RHD	Jean Richaud
Rmk	RMK	C.L.C. Rumker
Roe	ROE	E.D. Roe
Rss	RSS	J.M. Rousseau
Rst	RST	R.A. Rossiter
Rus	RUS	Henry Norris Russell
S	**S**	**James South**
San	SAN	Roscoe Sanford
Sbk	SBK	G.M. Seabroke
Sca	SCA	Marco Scardia
Scj	SCJ	H.C.F.C. Schjellerup
Sct	SCT	J.L. Scott
Se	SE	A. Secchi
See	SEE	T.J.J. See
Sei	SEI	J. Scheiner
Shb	SHB	J.M. Schaeberle
Sh	**SHJ**	**J. South & J. Herschel joint catalogue**
Sle	SLE	G. Soulie
Slo	SLO	F. Slocum
Slr	SLR	R.P. Sellors
Sma	SMA	W.M. Smart
Sml	SMK	Paul Schmidtke
Smy	**SMY**	**William Henry Smyth**
Sod	SOD	S. Soderhjelm
Sp	SP	Giovanni Schiaparelli
SprO	SPR	Sproul Observatory
St	ST	Carl Stearns
Σ	**STF**	**Wilhelm Struve – Dorpat Obs. catalogue**
Σ	**STF**	**Wilhelm Struve – 1st supplement**
Σ	**STF**	**Wilhelm Struve – 2nd supplement**
$G\Sigma$	STG	G. Struve
$H\Sigma$	STH	Hermann Struve
Sti	STI	John Stein
Stm	STM	Mark Stauffer
Stn	STN	Ormond Stone
Str	STR	K.A. Strand
$O\Sigma$	**STT**	**Otto Struve – Pulkovo Obs. catalogue**
$O\Sigma\Sigma$	**STT**	**O. Struve – Pulkovo Obs. cat. supplement**
Stu	STU	K. Sturdy
Swi	SWI	Lewis Swift
SydO	SYO	Sydney Observatory
Tar	TAR	K.J. Tarrant
Tay	TAY	P.H. Taylor
TDS	TDS	Tycho Double Star
Tea	TEA	E.T.H. Teague

Table A2. Codes for Star Designations *(continued)*

Designation	IDS/WDS Code	Discoverer or Observatory
Tgy	TGY	Ronald Charles Tanguay
Thk	THK	Ronald Thorkildson
Tob	TOB	Tofol Tobal
Tp	TP	N. Tapia
Tru	TRU	R.J. Trumpler
Tuc	TUC	Richard Tucker
TycC	TYC	Tycho-2 Catalogue 2000
vab	VAB	G.B. van Albada
VatO	VAT	Vatican Observatory
VBs	VBS	George van Biesbroeck
Vou	VOU	J.G.E.G. Voute
VanO	VVO	Van Vleck Observatory
Wak	WAK	R.L. Walker
Ward	WAR	I.W. Ward
War	WAR	O.C.R. Warren
Wat	WAT	R. Waterworth
WDF	WDF	Washington Fundamental Cat. (transits)
Webb	**WEB**	**T.W. Webb**
Wei	WEI	M. Weisse
Wey	WEY	C. Weymouth
West	WJD	J.D. West
WFC	WFC	Washington Fundamental Cat. (astrographs)
Wg	WG	R.W. Wrigley
Whc	WHC	H.C. Wilson
Wolf	WLF	Max Wolf
Wnc	WNC	F.A. Winnecke
WshO	WNO	U.S. Naval Observatory
Wor	WOR	Charles Worley
Wre	WRE	R.E. Wilson
Wrh	WRH	R.H. Wilson, Jr.
Ws	WS	J.M. Wilson
WSI	WSI	Washington Speckle Interferometry
Wz	WZ	Carl Wirtz
Y	Y	Yale (Observatory) Catalogue
Yng	YNG	C.A. Young
You	YOU	S.P. Young
YR	YR	Yale-Rochester (observatories)
YSJ	YSJ	Yale-San Juan (observatories)
Zag	ZAG	F. Zager
Zin	ZIN	E. Zinner
Zul	ZUL	D.J. Zulevic

NOTE: Another double star designation that is often seen in catalogues and observing lists is "ADS" – which stands for "Aitken Double Star" – followed by the running number in R.G. Aitken's great *New General Catalogue of Double Stars Within 120° of the North Pole*, compiled at the Lick Observatory in 1932. This monumental work was the predecessor to the *Index Catalogue of Visual Double Stars (IDS)* and included many discoveries by other observers in addition to his own (which carry the symbol "A" given above).

Appendix 3

Double and Multiple Star Working List

Presented here is an extended table of 400 double and multiple star systems intended for those who desire to see more of these starry jewels after viewing the hundred showpieces provided in Chapter 7, or who wish to pursue some of the projects suggested in Chapter 6, such as revising Dawes' Limit or making micrometer measures of binaries. This working list is arranged by Right Ascension rather than by constellation as in the showpiece roster. A few entries from that compilation will be found repeated here if one of their components happens to be a close binary of interest. A wide range of objects is offered, from naked-eye/binocular pairs to those requiring a 14-inch telescope and excellent seeing. Three objects of special interest lying below our –45-degree Declination limit are included.

Primary data sources for both lists were *Sky Catalogue 2000.0* and the *Washington Double Star Catalog*. Right Ascension (RA) in hours and minutes and Declination (Dec) in degrees and minutes are for the current standard Epoch 2000.0. Constellation (Con) abbreviations are the official three-letter designations adopted by the International Astronomical Union (see the constellation listing in Appendix 1.) Other table headings are the apparent visual magnitudes (Mags) of the components, their approximate current angular separation (Sep) in arc-seconds and their spectral types (Spec) on either the standard MKK (Morgan-Keenan-Kellman) system or the HD (Henry Draper) system, if available. (For more information on spectral classes see Michael Inglis's excellent *Observer's Guide to Stellar Evolution*, Springer-Verlag.) Position angles are not given for a variety of reasons (among them the confusion resulting from the common use of star diagonals with refracting and compound telescopes, producing "inside-out" mirror-images of the sky). Those observers desiring the latest available position angles, as well as measures of component separations, should consult the U.S. Naval Observatory's *Washington Double Star Catalog* on-line at http://ad.usno.navy.mil/wds/

Approximate distance in lightyears (LY) is also given in many cases. Unless an orbital period is indicated, or a pair is noted as being "optical" (meaning it consists of two unrelated stars that happen to lie along the same line of sight), the objects are common proper motion (or CPM) systems – those drifting through space together and, therefore, gravitationally-bound. In most (if not all) cases such pairs are actually very slowly orbiting each other, but in periods measured in thousands of years. Finally, this listing extends down to –45 degrees Declination, covering that three-fourths of the entire heavens visible from mid-northern latitudes.

Table A3. Double and Multiple Star Working List

Object/Con	RA	Dec	Mags	Sep	Spec	Remarks
WZ Cas	00ʰ 01ᵐ	+60° 21'	7.6–10, 8.7	58"	N1, A	Dim but striking; red and blue!
Σ3053 Cas	00ʰ 03ᵐ	+66° 06'	5.9, 7.3	15"	G0, A2	Beautiful orange and blue pair.
Σ2 Cep	00ʰ 09ᵐ	+79° 43'	6.6, 6.9	0.8"	A7IV	Tight 300-yr. binary.
κ-1 Scl	00ʰ 09ᵐ	−27° 59'	6.1, 6.2	1.4"	F2, F2	Slow (very long period) binary.
34 Psc	00ʰ 10ᵐ	+11° 09'	5.4, 9.4	8"	B8	Close, unequal pair.
OΣ2 And	00ʰ 13ᵐ	+26° 59'	6.7, 7.5	0.5"	G0III, F2IV	Binary – period 695 yrs.
35 Psc	00ʰ 15ᵐ	+08° 49'	6.0, 7.6	12"	F0IV, A7	Fixed (no orbital motion).
Σ13 Cep	00ʰ 16ᵐ	+76° 57'	7.0, 7.3	0.9"	B8V	Slow 1600-yr. binary.
Σ24 And	00ʰ 18ᵐ	+26° 08'	7.6, 8.4	5"	A2	Neat little pair.
λ Cas	00ʰ 32ᵐ	+54° 31'	5.5, 5.8	0.6"	B8V	Tight binary – period 640 yrs.
π And	00ʰ 37ᵐ	+33° 43'	4.4, 8.6	36"	B3	Wide, unequal double.
β395 Cet	00ʰ 37ᵐ	−24° 46'	6.3, 6.4	0.7"	G5V	Fast binary – period 25 yrs!
α Cas	00ʰ 40ᵐ	+56° 32'	2.2, 8.9	64"	K0	Mag. contrast pair – optical.
55 Psc	00ʰ 40ᵐ	+21° 26'	5.4, 8.7	6"	K0I, F3V	Fixed, orange and blue pair.
HN122 Cas	00ʰ 46ᵐ	+74° 59'	5.7–6.1, 9.4	36"	A2	Optical. Primary = YZ Cas.
h3395 Phe	00ʰ 46ᵐ	−41° 55'	8.4, 8.9	6"	K0	Faint reddish near-twins.
65 Psc	00ʰ 50ᵐ	+27° 43'	6.3, 6.3	4"	F4III, F5III	Identical twin yellowish duo.
36 And	00ʰ 55ᵐ	+23° 38'	6.0, 6.4	0.9"	K1IV	Binary – period 165 yrs.
66 Psc	00ʰ 55ᵐ	+19° 11'	6.2, 6.9	0.5"	A1V	Binary – period 360 yrs.
26 Cet	01ʰ 04ᵐ	+01° 22'	6.2, 8.6	16"	F0	Subtle color contrast.
77 Psc	01ʰ 06ᵐ	+04° 55'	6.8, 7.6	33"	F2, F2	Neat roomy pair.
β Phe	01ʰ 06ᵐ	−46° 43'	4.0, 4.3	1.4"	G8III	Bright tight pair, slow binary.

Table A3. Double and Multiple Star Working List (continued)

Object/Con	RA	Dec	Mags	Sep	Spec	Remarks
φ And	01ʰ 10ᵐ	+47° 15'	4.6, 5.5	0.5"	B7V	Very close 370-yr. binary.
β And	01ʰ 10ᵐ	+35° 37'	2.1, 11.8	80"	M	Galaxy NGC 404 in field.
φ Cas	01ʰ 20ᵐ	+58° 14'	5.1, 7.8	134"	F5, B5	In cluster NGC 457.
42 Cet	01ʰ 20ᵐ	−00° 31'	6.5, 6.8	2"	A7V	Slow binary.
ψ Cas	01ʰ 26ᵐ	+68° 08'	4.7, 9.6, 9.7	25", 3"	K0	Delicate triple – B-C fixed.
τ Scl	01ʰ 36ᵐ	−29° 54'	6.0, 7.1	2.3"	F4	Binary – period 1900 yrs.
ε Scl	01ʰ 46ᵐ	−25° 03'	5.4, 8.6	5"	F0	Binary – period 1200 yrs.
1 Ari	01ʰ 50ᵐ	+22° 17'	6.2, 7.4	3"	K1III, A6V	Slow binary.
χ Cet	01ʰ 50ᵐ	−10° 41'	4.9, 6.9	184"	F3III, G0	Wide, bright easy pair.
Σ163 Cas	01ʰ 51ᵐ	+64° 51'	6.8, 8.8	35"	K5	Pretty orange and blue pair.
ζ Cet	01ʰ 52ᵐ	−10° 20'	3.7, 9.9	187"	K0, K0	Wide orange mag. contraster.
Σ186 Cet	01ʰ 56ᵐ	+01° 51'	6.8, 6.8	1.1"	F9V	Identical-twin 170-yr. binary.
48 Cas	02ʰ 02ᵐ	+70° 54'	4.7, 6.4	0.9"	A3IV	Binary – period 60 yrs.
γ-2 And	02ʰ 04ᵐ	+42° 20'	5.5, 6.3	0.4"	B9V, A0V	Blue and green, 61-yr. binary.
10 Ari	02ʰ 04ᵐ	+25° 56'	5.9, 7.3	1.3"	F8IV	Binary – period 309 yrs.
59 And	02ʰ 11ᵐ	+39° 02'	6.1, 6.8	17"	A0, A2	Neat fixed bluish-white pair.
66 Cet	02ʰ 13ᵐ	−02° 24'	5.7, 7.5	16"	F8V	Slow binary, yellow and blue.
Σ239 Tri	02ʰ 17ᵐ	+28° 45'	7.0, 8.0	14"	F5	Neat slivery-white pair.
ω For	02ʰ 34ᵐ	−28° 14'	5.0, 7.7	11"	B9V	Slow binary.
15 Tri	02ʰ 36ᵐ	+34° 41'	5.7, 6.9	140"	M, A5	Wide color contraster.
30 Ari	02ʰ 37ᵐ	+24° 39'	6.6, 7.4	39"	F5V, F6III	Easy wide yellowish pair.
OΣ44 Per	02ʰ 42ᵐ	+42° 47'	8.4, 9.1	1.4"	B9	In cluster M34.

Table A3. Double and Multiple Star Working List *(continued)*

Object/Con	RA	Dec	Mags	Sep	Spec	Remarks
h1123 Per	02ʰ 42ᵐ	+42° 47'	8.0, 8.0	20"	A0, A0	In cluster M34.
θ Per	02ʰ 44ᵐ	+49° 14'	4.1, 9.9	20"	F7V, M1V	Unequal, wide slow binary.
Σ305 Ari	02ʰ 48ᵐ	+19° 22'	7.4, 8.2	4"	F9V	Binary – period 720 yrs.
π Ari	02ʰ 49ᵐ	+17° 28'	5.2, 8.7, 10.8	3", 25"	B6V	Challenging triple.
ε Ari	02ʰ 59ᵐ	+21° 20'	5.2, 5.5	1.4"	A2V, A2V	Close, matched slow binary.
Σ331 Per	03ʰ 01ᵐ	+52° 21'	5.3, 6.7	12"	B5	Nice easy double.
α+93 Cet	03ʰ 02ᵐ	+04° 05'	2.5, 5.6	960"	M2III, B7III	Ultra-wide red and blue duo.
α For	03ʰ 12ᵐ	−28° 59'	4.0, 7.0	5"	F7IV, G7V	Binary – period 314 yrs.
95 Cet	03ʰ 18ᵐ	−00° 56'	5.6, 7.5	1.0"	K1IV, G8V	Binary – period 217 yrs.
τ-4 Eri	03ʰ 20ᵐ	−21° 45'	3.7, 9.2	6"	M2	Tight mag. contrast pair.
7 Tau	03ʰ 34ᵐ	+24° 28'	6.6, 6.7	0.7"	A3V	Close 568-yr. binary.
Σ400 And	03ʰ 35ᵐ	+60° 02'	6.8, 7.6	1.6"	F4V	Binary – period 288 yrs.
Σ422 Tau	03ʰ 37ᵐ	+00° 35'	5.9, 8.8	7"	G9V, K6V	With 10 Tau in field.
η Tau	03ʰ 48ᵐ	+24° 06'	2.9, 8.0 8.0, 8.6	117" 180", 190"	B5, A0 A0, G0	Striking, delicate quadruple system in Pleiades Cluster.
30 Tau	03ʰ 48ᵐ	+11° 09'	5.1, 10.1	9"	B3	Tough mag. contrast pair.
f Eri	03ʰ 49ᵐ	−37° 37'	4.8, 5.3	8"	B8, A0	Lovely bright double.
ζ Per	03ʰ 54ᵐ	+31° 53'	2.9, 9.5, 9.5	13", 4"	B1I	Fixed. Other stars close by.
OΣ67 Cam	03ʰ 57ᵐ	+61° 07'	5.3, 8.5	1.9"	K3II	Fixed, gold and green pair.
ε Per	03ʰ 58ᵐ	+40° 01'	2.9, 7.6	9"	B0V, A2V	Like ζ Per. Fixed.
Σ484 Cam	04ʰ 07ᵐ	+62° 23'	10, 10, 10	5", 23"	—	In open cluster NGC 1502.
Σ485 Cam	04ʰ 08ᵐ	+62° 20'	7.0, 7.1, 9.8	18", 70"	B0	In open cluster NGC 1502.
Σ460 Cep	04ʰ 10ᵐ	+80° 42'	5.5, 6.3	0.6"	G8III, A6V	Binary – period 415 yrs.

Table A3. Double and Multiple Star Working List *(continued)*

Object/Con	RA	Dec	Mags	Sep	Spec	Remarks
39 Eri	04h 14m	−10° 15'	5.0, 8.0	6"	K3III	Distant 9.5-mag. star.
φ Tau	04h 20m	+27° 21'	5.0, 8.4	52"	K0	Wide optical pair.
β87 Tau	04h 22m	+20° 49'	6.0, 9.1	1.9"	M0, A0	Close, dim red and blue duo.
χ Tau	04h 23m	+25° 38'	5.5, 7.6	19"	B9	Fixed pair.
Σ552 Per	04h 31m	+40° 01'	7.0, 7.2	9"	B8	Neat matched combo.
81 Tau	04h 31m	+15° 42'	5.5, 9.4	162"	A5, K0	Wide color/mag. contraster.
1 Cam	04h 32m	+53° 55'	5.7, 6.8	10"	B0III	Attractive but neglected pair.
57 Per	04h 33m	+43° 04'	6.1, 6.8	116"	F0, F0	Nice wide matched duo.
α Tau	04h 36m	+16° 31'	0.8–1.0, 11	122"	K5III	Radiant Aldebaran! Optical.
53 Eri	04h 38m	−14° 18'	4.0, 7.0	0.7"	K0	Tight bright pair – closing.
2 Cam	04h 40m	+53° 28'	5.6, 7.3	0.7"	F5V	Binary – period 425 yrs.
55 Eri	04h 44m	−08° 48'	6.7, 6.8	9"	G8III, F4III	Pretty matched twins.
σ-1/2 Tau	04h 39m	+15° 55'	4.7, 5.1	430"	A3, A2	Wide pair in Hyades cluster.
ω Aur	04h 59m	+37° 53'	5.0, 8.0	5"	A0	Slow, tight binary.
β314 Lep	04h 59m	−16° 23'	5.9, 7.3	0.9"	F3V, F9V	Binary – period 55 yrs.
Σ627 Ori	05h 01m	+03° 37'	6.6, 7.0	21"	A0, A0	Neat nearly matched duo.
β Cam	05h 03m	+60° 27'	4.0, 8.6	81"	G0, A5	Roomy mag. contrast pair.
γ Cae	05h 04m	−35° 29'	4.6, 8.1	3"	K0	Tight mag. contrast pair.
11/12 Cam	05h 06m	+58° 58'	5.2, 6.1	179"	B2V, K0III	Wide, striking color contrast.
14 Ori	05h 08m	+08° 30'	5.8, 6.5	0.8"	A0	Binary – period 200 yrs.
Σ644 Aur	05h 10m	+37° 18'	6.7, 7.0	1.6"	B2II, K3	Lovely tight color contraster.
ρ Ori	05h 13m	+02° 52'	4.5, 8.3	7"	K0	Fixed tight mag. contrast duo.

Table A3. Double and Multiple Star Working List *(continued)*

Object/Con	RA	Dec	Mags	Sep	Spec	Remarks
κ Lep	05^h 13^m	−12° 56'	4.5, 7.4	2.6"	B8	Tight mag. contrast pair.
14 Aur	05^h 15^m	+32° 41'	5.1, 7.4–7.9	15"	A9IV, A2	Neat pair with var. comp.
S476 Lep	05^h 19^m	−18° 31'	6.2, 6.4	39"	B8, B8	Matched bluish pair.
h3750 Lep	05^h 20^m	−21° 14'	4.7, 8.5	4"	A0	Pretty mag. contrast pair.
Σ681 Aur	05^h 21^m	+46° 58'	6.7, 8.7	23"	F0	In wide field with Capella.
22 Ori	05^h 22^m	−00° 23'	4.7, 5.7	242"	B2IV, B3V	Very wide blue-white combo.
41 Lep	05^h 22^m	−24° 46'	5.4, 6.6, 9.1	3", 62"	G0, A3, K0	Unequal tinted triple.
23 Ori	05^h 23^m	+03° 33'	5.0, 7.1	32"	O9II, B2V	Nice blue-white, easy pair.
Σ698 Aur	05^h 25^m	+34° 51'	6.6, 8.7	31"	K0, K	Attractive orange combo.
β Lep	05^h 28^m	−20° 46'	2.8, 7.3	2.5"	G0	Bright close, unequal double.
118 Tau	05^h 29^m	+25° 09'	5.8, 6.6	5"	B8V, A1V	Neat snug, blue-white pair.
32 Ori	05^h 31^m	+05° 57'	4.5, 5.8	1.1"	B5IV	Bright, close 585-yr. binary
33 Ori	05^h 31^m	+03° 18'	5.8, 7.1	1.8"	B3	Snug unequal pair.
α Lep	05^h 33^m	−17° 49'	2.6, 11, 12	36", 91"	F0	Wide, dim delicate triple.
Σ750 Ori	05^h 35^m	−04° 22'	6.5, 8.5	4"	B5	In cluster NGC 1981.
Σ743 Ori	05^h 35^m	−04° 24'	8.3, 9.4	2"	B8	In cluster NGC 1981.
θ-2 Ori	05^h 35^m	−05° 25'	5.2, 6.6	52"	O9V, B7IV	Wide pair in Orion Nebula.
Σ737 Aur	05^h 36^m	+34° 08'	8.5, 9.0	11"	B	In open cluster M36.
Σ742 Tau	05^h 36^m	+22° 00'	7.2, 7.8	4"	F8	Neat duo near Crab Nebula.
26 Aur	05^h 39^m	+30° 30'	6.0, 8.0	12"	A2	Yellow and blue combo.
σ Ori AB	05^h 39^m	−02° 36'	4.0, 6.0	0.2"	B0	Ultra-tight 170-yr. binary.
h3780 Lep	05^h 39^m	−17° 51'	7.5, 8.5, 8.4	89", 76"	B9	Other stars = cl. NGC 2017.

Table A3. Double and Multiple Star Working List *(continued)*

Object/Con	RA	Dec	Mags	Sep	Spec	Remarks
52 Ori	$05^h 48^m$	$+06° 27'$	5.3, 5.3	1.2"	A3	Perfect twins – in contact!
α Ori	$05^h 55^m$	$+07° 24'$	0.4–1.3, 10.6	174"	M1-M2I	Radiant Betelgeuse! Optical. Plus 9th-mag. at 118".
Σ855 Ori	$06^h 09^m$	$+02° 30'$	6.0, 7.0	29"	A0	Neat slow, white binary.
41 Aur	$06^h 12^m$	$+48° 43'$	6.3, 7.0	8"	A0, A0	Challenging 474-yr. binary.
η Gem	$06^h 15^m$	$+22° 30'$	3.3–3.9, 8.8	1.6"	M3III	Nice color contrast.
Σ872 Aur	$06^h 16^m$	$+36° 09'$	6.9, 7.9	11"	F0	Bright, wide tinted duo.
ζ CMa	$06^h 20^m$	$-30° 04'$	3.0, 7.6	176"	B8, K0	Triple in cluster NGC 2232.
10 Mon	$06^h 28^m$	$-04° 46'$	5.1, 9.3, 9.3	77", 81"	B3	Wide unequal pair.
ν Gem	$06^h 29^m$	$+20° 13'$	4.2, 8.7	112"	B5	Yellow and blue fixed pair.
20 Gem	$06^h 32^m$	$+17° 47'$	6.3, 6.9	20"	F8III	Snug matched double.
β755 Col	$06^h 35^m$	$-36° 47'$	6.0, 6.8	1.3"	B9	Dim, nearly equilateral triple.
Σ939 Mon	$06^h 36^m$	$+05° 18'$	8.3, 9.6, 9.7	30", 40"	B5, B8	Fixed mag. contrast pair.
ν-1 CMa	$06^h 36^m$	$-18° 40'$	5.8, 8.5	18"	G5, G0	In open cluster NGC 2264.
15/S Mon	$06^h 41^m$	$+09° 53'$	3.9, 7.4, 7.7	3", 156"	O5	Close 480-yr. binary.
14 Lyn	$06^h 53^m$	$+59° 27'$	5.6, 6.8	0.4"	G0I, A2	Slow binary – color contrast.
38 Gem	$06^h 55^m$	$+13° 11'$	4.7, 7.7	7"	F0V, G4V	Tight orange and blue – fixed.
μ CMa	$06^h 56^m$	$-14° 03'$	5.3, 8.6	3"	G5	Bright, tight slow binary.
15 Lyn	$06^h 57^m$	$+59° 25'$	4.8, 5.9	0.9"	G0	Adhara. Like fainter Sirius!
ε CMa	$06^h 59^m$	$-28° 58'$	1.5, 7.8	8"	B2II, B6	Nice pair + wide orange star.
δ38 Pup	$07^h 04^m$	$-43° 36'$	5.6, 7.2, 8.1	20", 185"	G0, G0, K2	Cozy, matched duo.
Σ1009 Lyn	$07^h 06^m$	$+52° 45'$	6.9, 7.0	4"	A2	Perfect twin yellowish pair.
Σ1035 Gem	$07^h 12^m$	$+22° 17'$	8.2, 8.2	9"	F5, F5	

Table A3. Double and Multiple Star Working List (continued)

Object/Con	RA	Dec	Mags	Sep	Spec	Remarks
π Pup	07ʰ 17ᵐ	−37° 06'	2.7, 8.0	69"	K5, B9	Wide color/mag. contraster.
λ Gem	07ʰ 18ᵐ	+16° 32'	3.6, 10.7	10"	A2	Dim, delicate mag. contrast.
τ CMa	07ʰ 19ᵐ	−24° 57'	4.4, 8.8	85"	O9I	Heart of cluster NGC 2362.
20 Lyn	07ʰ 22ᵐ	+50° 09'	7.3, 7.4	15"	F0, F0	Nice matched pair.
19 Lyn	07ʰ 23ᵐ	+55° 17'	5.6, 6.5	15"	B8V, A0V	Neat duo with 8.9-mag. near.
η CMa	07ʰ 24ᵐ	−29° 18'	2.4, 6.9	179"	B7, A0	Bright wide – color contrast.
Σ1104 Pup	07ʰ 29ᵐ	−15° 00'	6.4, 7.5	2"	F7V	Binary – period 1100 yrs.
σ Pup	07ʰ 29ᵐ	−43° 18'	3.3, 9.4	22"	M0, G5	Color/mag. contrast pair.
h3973 Pup	07ʰ 32ᵐ	−20° 56'	8.3, 9.3	9"	B8	Dim white and red pair.
Σ1108 Gem	07ʰ 33ᵐ	+22° 53'	6.5, 8.3	12"	G5	Easy unequal double.
η Pup	07ʰ 34ᵐ	−23° 28'	5.1, 5.1	10"	F4, F5	Striking twins – slow binary.
Σ1121 Pup	07ʰ 37ᵐ	−14° 30'	7.9, 7.9	7"	B9, B9	Equal pair – in cluster M47.
Σ1126 Cmi	07ʰ 40ᵐ	+05° 14'	6.6, 6.9	0.9"	A0	In field with Procyon.
κ Gem	07ʰ 44ᵐ	+24° 24'	3.6, 8.1	7"	G8III	Mag. contrast – slow binary.
β Gem	07ʰ 45ᵐ	+28° 02'	1.1, 8.8, 9.6	201", 234"	K0III	Pollux. Other fainter comps.
2 Pup	07ʰ 46ᵐ	−14° 41'	6.1, 6.8	17"	A0, A0	Neat nearly matched pair.
9 Pup	07ʰ 52ᵐ	−13° 54'	5.6, 6.2	0.6"	G1V	Ultra-close 23-yr. binary.
14 CMi	07ʰ 58ᵐ	+02° 13'	5.4, 8.4, 9.3	89", 120"	K0	Wide delicate triple.
ζ Mon	08ʰ 09ᵐ	−02° 59'	4.3, 7.8	66"	G2, K2	Nice wide optical pair.
h4063 Pup	08ʰ 16ᵐ	−37° 22'	7.5, 9.6	18"	B8	Blue-white and red pair.
Σ1216 Hya	08ʰ 21ᵐ	−01° 36'	7.1, 7.4	0.7"	A2V	Binary – period 435 yrs.
φ-2 Cnc	08ʰ 27ᵐ	+26° 56'	6.3, 6.3	5"	A6V, A3V	Identical twin slow binary.

Table A3. Double and Multiple Star Working List *(continued)*

Object/Con	RA	Dec	Mags	Sep	Spec	Remarks
24 Cnc	08h 27m	+24° 32'	7.0, 7.8	6"	F7V, G	B unresolved 22-yr. binary.
β205 Pyx	08h 33m	−24° 36'	6.9, 7.0	0.6"	A8IV	Matched 160-yr. binary.
Σ1245 Cnc	08h 36m	+06° 37'	6.0, 7.2	10"	F6, G5	Two wider, fainter comps.
β208 Pyx	08h 39m	−22° 40'	5.3, 6.7	1.1"	G6	Binary – period 145 yrs.
S571 Cnc	08h 40m	+19° 33'	6.9, 7.2, 6.7	45", 93"	A0, A0, K0	Triple in Beehive Cluster.
39 Cnc	08h 40m	+20° 00'	6.5, 6.5	150"	K0, K0	Wide orange pair in Beehive.
ε Cnc	08h 40m	+19° 33'	6.3, 7.4	135"	A2, A0	Another wide one in Beehive.
Σ1254 Cnc	08h 40m	+19° 40'	6.4, 8.9 8.6, 8.9	20" 63", 83"	G5, A0 —	Wide, delicate quadruple system in Beehive Cluster.
ζ Pyx	08h 40m	−29° 34'	4.9, 9.1	52"	G4, G0	Unequal mag. contrast pair.
γ Cnc	08h 43m	+21° 28'	4.7, 8.7	106"	A0	Wide mag. contrast combo.
57 Cnc	08h 54m	+30° 35'	6.0, 6.5, 9.1	1.4", 56"	G7III, K0	Close pair both yellow.
17 Hya	08h 56m	−07° 58'	6.8, 7.0	4"	A3	Neat cozy matched duo.
10 UMa	09h 01m	+41° 47'	4.1, 6.2	0.6"	F5V	Bright, rapid 22-yr. binary.
σ-2 UMa	09h 10m	+67° 08'	4.8, 8.2, 9.3	4", 205"	F7IV	Close pair 1100-yr. binary.
Σ1321 UMa	09h 14m	+52° 41'	7.6, 7.7	17"	M0V, M0V	Wide ruddy 975-yr. binary.
27 Hya	09h 20m	−09° 33'	5.0, 6.9, 9.1	229", 9"	G5, F2	Unequal triple.
Σ1338 Lyn	09h 21m	+38° 11'	6.5, 6.7	0.5"	F3V	Binary – period 220 yrs.
κ Leo	09h 25m	+26° 11'	4.5, 10, 10	3", 53"	K0	Challenging triple.
ω Leo	09h 28m	+09° 03'	5.9, 6.5	0.6"	F9V	Binary – period 118 yrs.
τ-1 Hya	09h 29m	−02° 46'	4.9, 7.9	66"	F5	Wide color contrast pair.
ζ-1 Ant	09h 31m	−31° 53'	6.2, 7.1	8"	A0	Fixed pair. ζ-2 near.

Table A3. Double and Multiple Star Working List *(continued)*

Object/Con	RA	Dec	Mags	Sep	Spec	Remarks
ψ Vel	09ʰ 31ᵐ	−40° 28'	4.1, 4.6	0.7"	F2IV	Bright rapid 34-yr. binary.
23 UMa	09ʰ 32ᵐ	+63° 04'	3.7, 8.9	23"	F0	Fixed mag. contrast pair.
Σ1365 HYA	09ʰ 32ᵐ	+01° 28'	7.8, 8.4	3"	F8	Subdued close double.
φ UMa	09ʰ 52ᵐ	+54° 04'	5.3, 5.4	0.3"	A3IV	Ultra-tight 106-yr. binary.
γ Sex	09ʰ 53ᵐ	−08° 06'	5.6, 6.1	0.6"	A1V	Rapid binary – period 76 yrs.
OΣ215 Leo	10ʰ 16ᵐ	+17° 44'	7.2, 7.4	1.5"	A9IV	Binary – period 550 yrs.
ζ+35 Leo	10ʰ 17ᵐ	+23° 25'	3.4, 5.9	330"	F0III, G2IV	Wide bright binocular pair.
49 Leo	10ʰ 35ᵐ	+08° 39'	5.8, 8.5	2.4"	A0	Like dim δ Cyg.
β411 Hya	10ʰ 36ᵐ	−26° 40'	6.7, 7.5	1.4"	F6V	Binary – period 210 yrs.
35 Sex	10ʰ 43ᵐ	−04° 45'	6.3, 7.4	7"	K0	Golden-orange and blue-green.
Σ1474 Hya	10ʰ 48ᵐ	−15° 16'	6.7, 7.8, 6.8	70", 76"	A0, —, F5	In field with Σ1473.
Σ1473 Hya	10ʰ 48ᵐ	−15° 37'	7.7, 8.6	31"	F8	In field with Σ1474.
Σ1495 UMa	11ʰ 00ᵐ	+58° 54'	7.2, 9.5	34"	K2	Gold and blue double.
α UMa	11ʰ 04ᵐ	+61° 45'	1.9, 4.8, 7.0	0.7", 378"	KOIII, —, F8	45-yr. binary + tinted combo.
ν UMa	11ʰ 19ᵐ	+33° 06'	3.5, 9.9	7"	K0	Fixed mag. contrast pair.
Σ1529 Leo	11ʰ 19ᵐ	−01° 39'	7.0, 8.0	10"	F8	Neat pair with subtle hues.
ι Leo	11ʰ 24ᵐ	+10° 32'	4.0, 6.7	1.8"	F2IV, F4	Binary – period 192 yrs.
τ Leo	11ʰ 28ᵐ	+02° 51'	5.1, 8.0	91"	K0, G5	Color and mag. contrast pair.
57 UMa	11ʰ 29ᵐ	+39° 20'	5.3, 8.3	5"	A2	Close unequal slow binary.
OΣ235 UMa	11ʰ 32ᵐ	+61° 05'	5.8, 7.1	0.7"	F6V	Rapid binary – period 73 yrs.
88 Leo	11ʰ 32ᵐ	+14° 22'	6.4, 8.4	15"	G0	Fixed unequal pair.
I78 Cen	11ʰ 34ᵐ	−40° 35'	6.2, 6.2	1.0"	A2	Snug matched fixed pair.

Table A3. Double and Multiple Star Working List *(continued)*

Object/Con	RA	Dec	Mags	Sep	Spec	Remarks
90 Leo	11ʰ 35ᵐ	+16° 48'	6.7, 7.3, 8.7	3", 63"	B3	Unequal triple system.
Σ1561 UMa	11ʰ 39ᵐ	+45° 07'	6.3, 8.4, 8.5	9", 85"	G0, —, F2	Also distant 9ᵗʰ-mag.
β Leo	11ʰ 49ᵐ	+14° 34'	2.1, 8.5	264"	A3V, F8	Denebola. 5.9-mag. in field.
β Hya	11ʰ 53ᵐ	−33° 54'	4.7, 5.5	0.9"	B9	Slow binary – closing.
65 UMa	11ʰ 55ᵐ	+46° 29'	6.7, 8.3, 6.5	4", 63"	A0	Triple system – like 90 Leo.
ζ Com	12ʰ 04ᵐ	+21° 28'	5.9, 7.4	4"	F0	Tight fixed pair.
2 Cvn	12ʰ 16ᵐ	+40° 40'	5.8, 8.1	11"	M1III, F7V	Golden-orange and blue duo.
Σ1627 Vir	12ʰ 18ᵐ	−03° 57'	6.6, 6.9	20"	F0, F0	Nice matched double.
Σ1633 Com	12ʰ 21ᵐ	+27° 03'	7.0, 7.1	9"	F2	Pretty matched pair.*
Wnc4 UMa	12ʰ 22ᵐ	+58° 05'	9.0, 9.3	50"	—	Wide dim pair = M40.
17 Vir	12ʰ 22ᵐ	+05° 18'	6.6, 9.4	20"	F8	Subtle color contrast.
Σ1639 Com	12ʰ 24ᵐ	+25° 35'	6.8, 7.8	1.7"	F0V	Binary – period 680 yrs.*
17 Com	12ʰ 29ᵐ	+25° 55'	5.3, 6.6	145"	A0, A3	* In Coma star cluster.
Σ1664 Vir	12ʰ 38ᵐ	−11° 31'	8.1, 9.3	26"	K0, G5	In Sombrero Galaxy field.
Σ1669 Crv	12ʰ 41ᵐ	−13° 01'	6.0, 6.1	5"	F5V, F3V	Lovely matched double.
γ Cen	12ʰ 42ᵐ	−48° 58'	2.9, 2.9	0.8"	A0III	Brilliant tight 84-yr. binary.
32/33 Com	12ʰ 52ᵐ	+17° 04'	6.3, 6.7	195"	M0III, F8	Wide, colorful binocular duo.
35 Com	12ʰ 53ᵐ	+21° 14'	5.1, 7.2, 9.1	1.2", 29"	G8III, F6, —	Close pair 500-yr. binary.
78 UMa	13ʰ 01ᵐ	+56° 22'	5.0, 7.4	1.5"	F2V	Neat, tight 115-yr. binary.
17 CVn	13ʰ 10ᵐ	+38° 30'	6.0, 6.2	84"	F0, B9	Wide matched pair.
α Com	13ʰ 10ᵐ	+17° 32'	5.0, 5.1	0.4"	F6V	Close, rapid 26-yr. binary.
θ Vir	13ʰ 10ᵐ	−05° 32'	4.4, 9.4, 10.4	7", 70"	A1V	Delicate fixed triple.

Table A3. Double and Multiple Star Working List *(continued)*

Object/Con	RA	Dec	Mags	Sep	Spec	Remarks
54 Vir	13ʰ 13ᵐ	−18° 50'	6.8, 7.3	5"	A0	Neat cozy pair.
OΣΣ123 Dra	13ʰ 27ᵐ	+64° 44'	6.7, 7.0	69"	F0, F0	Nice matched wide duo.
β932 Vir	13ʰ 35ᵐ	−13° 13'	6.5, 6.9	0.4"	A0V	Very close 200-yr. binary.
25 CVn	13ʰ 38ᵐ	+36° 18'	5.0, 6.9	1.8"	A7III	Binary – period 240 yrs.
1 Boo	13ʰ 41ᵐ	+19° 57'	5.8, 8.7	5"	A2	Tight mag. contrast pair.
Σ1785 Boo	13ʰ 49ᵐ	+26° 59'	7.6, 8.0	3"	N2	Reddish 155-yr. binary.
4 Cen	13ʰ 52ᵐ	−31° 56'	4.7, 8.4	15"	B7	Fixed mag. contrast pair.
3 Cen	13ʰ 52ᵐ	−33° 00'	4.5, 6.0	8"	B5, B8	Striking bright fixed pair.
η Boo	13ʰ 55ᵐ	+18° 24'	2.7, 8.7	112"	G0	Attractive wide unequal duo.
Σ1788 Vir	13ʰ 55ᵐ	−08° 04'	6.5, 7.7	3.5"	F8V	Slow binary.
ι Boo	14ʰ 16ᵐ	+51° 22'	4.9, 7.5	38"	A5	Nice fixed double.
Σ1835 Boo	14ʰ 23ᵐ	+08° 27'	5.1, 7.4	6"	A0V, F3V	Neat cozy pair.
τ-1 Lup	14ʰ 26ᵐ	−45° 13'	4.6, 9.3	158"	B3, M0	Wide color/mag. contraster.
Σ1838 Boo	14ʰ 24ᵐ	+11° 15'	7.4, 7.5	9"	F5	Pretty identical twins.
φ Vir	14ʰ 28ᵐ	−02° 14'	4.8, 9.3	5"	K0	Challenging mag. contraster.
Σ1850 Boo	14ʰ 29ᵐ	+28° 17'	7.0, 7.4	26"	A0, A0	Easy matched combo.
54 Hya	14ʰ 46ᵐ	−25° 27'	5.1, 7.1	9"	F2III, F9	Nice pair with subtle tints.
Σ1883 Vir	14ʰ 49ᵐ	+05° 57'	7.6, 7.6	0.9"	F6V	Twin binary – period 228 yrs.
μ Lib	14ʰ 49ᵐ	−14° 09'	5.8, 6.7	1.8"	A2	Tight slow binary.
39 Boo	14ʰ 50ᵐ	+48° 43'	6.2, 6.9	3"	F6V, F5V	Slow binary, closing.
α Lib	14ʰ 51ᵐ	−16° 02'	2.8, 5.2	230"	A3IV, F4IV	Bright binocular double.
HN28 Lib	14ʰ 57ᵐ	−21° 25'	5.7, 8.0	23"	K4V, M0	Orange and ruddy combo.

Appendices

Table A3. Double and Multiple Star Working List *(continued)*

Object/Con	RA	Dec	Mags	Sep	Spec	Remarks
59 Hya	14h 59m	−27° 39'	6.3, 6.6	0.3"	A4V	Ultra-close 340-yr. binary.
44 Boo	15h 04m	+47° 39'	5.3, 5.8–6.4	2"	G0V, G2	Binary – period 220 yrs.
π Lup	15h 05m	−47° 03'	4.6, 4.7	1.4"	B5, B5	Bright slow binary – opening.
ι Lib	15h 12m	−19° 47'	5.1, 9.4	58"	B9IV	Wide mag. contrast pair.
δ Boo	15h 16m	+33° 19'	3.5, 7.4	105"	G8III, G0V	Wide yellow and bluish pair.
Σ1932 CrB	15h 18m	+26° 50'	7.3, 7.4	1.6"	G0V	Snug matched 203-yr. binary.
5 Ser	15h 19m	+01° 46'	5.1, 10	11"	G0	Fixed pair, near globular M5.
η CrB	15h 23m	+30° 17'	5.6, 5.9	1.0"	G2V, G2	Rapid binary – period 42 yrs.
ε Lup	15h 23m	−44° 41'	3.7, 5.2, 8.8	0.6", 26"	B3	A-B likely binary.
π-1 UMi	15h 29m	+80° 27'	6.6. 7.3	31"	G5	Neat, far-northern pair.
ν-1/2 Boo	15h 31m	+40° 50'	5.0, 5.0	900"	K5III, A5V	Binoc. pair – orange and blue.
γ Lup	15h 35m	−41° 10'	3.5, 3.6	0.7"	B3V	Bright, tight 147-yr. binary.
Σ1964 CrB	15h 38m	+36° 15'	7.0, 7.6, 8.7	15", 16"	F5	Neat equilateral triangle.
Σ1962 Lib	15h 39m	−08° 47'	6.5, 6.6	12"	F6V, F6V	Striking, identical twin suns.
γ CrB	15h 43m	+26° 18'	4.1, 5.5	0.7"	A0IV	Close binary – period 91 yrs.
2 Sco	15h 54m	−25° 20'	4.7, 7.4	2"	B3	Snug unequal pair.
ξ Lup	15h 57m	−33° 58'	5.3, 5.8	10"	A3V, B9V	Striking matched fixed pair.
η Lup	16h 00m	−38° 24'	3.6, 7.8, 9.3	15", 115"	B3, –, F5	Fixed unequal triple.
ξ Sco AB	16h 04m	−11° 22'	4.8, 5.1	0.7"	F5IV	Rapid binary – period 46 yrs.
ω-1/2 Sco	16h 07m	−20° 40'	4.0, 4.3	720"	B1V, G3III	Naked-eye/binocular combo.
12 Sco	16h 12m	−28° 25'	5.9, 7.9	4"	B9	Cozy unequal pair.
Σ2021 Her	16h 13m	+13° 32'	7.4, 7.5	4"	K0	Very slow matched binary.

Table A3. Double and Multiple Star Working List *(continued)*

Object/Con	RA	Dec	Mags	Sep	Spec	Remarks
σ Sco	16h 21m	−25° 36'	2.9, 8.5	20"	B1, B9	Nice mag. contrast.
ν-1/2 CrB	16h 22m	+33° 48'	5.2, 5.4	360"	M2III, K5III	Wide, colorful binocular pair.
Σ2054 Dra	16h 24m	+61° 42'	6.0, 7.2	1.0"	G5	In field with η Dra.
η Dra	16h 24m	+61° 31'	2.7, 8.7	5"	G5	Tight unequal, slow binary.
ρ Oph	16h 26m	−23° 27'	5.3, 6.0	3"	B2IV, B2V	Slow binary. 7.0, 7.9 nearby.
Σ2052 Her	16h 29m	+18° 25'	7.7, 7.8	2"	K2V	Snug 236-yr. binary.
λ Oph	16h 31m	+01° 59'	4.2, 5.3	1.5"	A2V	Nice, cozy 130-yr. binary.
37 Her	16h 41m	+04° 13'	5.8, 7.0	70"	A0, A0	Neat wide pair.
ζ Her	16h 41m	+31° 36'	2.9, 5.5	0.7"	F9IV, G7V	Bright, rapid 34-yr. binary.
See293 Sco	16h 54m	−41° 48'	5.6, 7.3	57"	B0, B0	In open cluster NGC 6231.
ζ-1/2 Sco	16h 54m	−42° 22'	3.6, 4.7	408"	K4III, B1I	Bright, ultra-wide and colorful.
μ-1/2 Sco	16h 52m	−38° 03'	3.0, 3.6	347"	B2V, B2IV	Bright, ultra-wide bluish duo.
20 Dra	16h 56m	+65° 02'	7.3, 7.3	1.4"	F2V	Twin binary – period 580 yrs.
24 Oph	16h 57m	−23° 09'	6.2, 6.5	0.8"	A0	Close nearly-equal pair.
Σ2120 Her	17h 05m	+28° 05'	7.3, 10.1	17"	K0	Dim tinted optical pair.
η Oph	17h 10m	−15° 43'	3.0, 3.5	0.6"	A2V	Bright close 84-yr. binary.
MlbO4 Sco	17h 19m	−34° 59'	6.1, 7.6	1.8"	K3V	Rapid 42-yr. binary.
ν Ser	17h 21m	−12° 51'	4.3, 8.3	48"	A0	Wide mag. contrast pair.
Σ2173 Oph	17h 30m	−01° 04'	6.0, 6.1	1.1"	G8IV	Binary – period 46 yrs.
λ+υ Sco	17h 34m	−37° 06'	1.6, 2.7	2100" (35')	B2IV, B2IV	The Stingers. Famed brilliant naked-eye/binocular combo.
26 Dra	17h 35m	+61° 52'	5.3, 8.0	1.6"	G0V	Unequal, close 76-yr. binary.

Table A3. Double and Multiple Star Working List (continued)

Object/Con	RA	Dec	Mags	Sep	Spec	Remarks
61 Oph	17ʰ 45ᵐ	+02° 35'	6.2, 6.6	21"	A1V, A0V	Neat matched white pair.
90 Her	17ʰ 53ᵐ	+40° 00'	5.2, 8.5	1.6"	K0	Slow binary – color contrast.
67 Oph	18ʰ 01ᵐ	+02° 56'	4.0, 8.6	54"	B5	Wide mag. contrast pair.
7/9 Sgr	18ʰ 03ᵐ	−24° 17'	5.3, 6.0	900"	F3III, O5	Wide pair in Lagoon Nebula.
τ Oph	18ʰ 03ᵐ	−08° 11'	5.2, 5.9	1.7"	F4IV, F3	Binary – period 280 yrs.
HN40 Sgr	18ʰ 03ᵐ	−23° 02'	7.6, 10.7, 8.7	6", 11"	O7	Triple in heart of Trifid Neb.
h5014 CrA	18ʰ 07ᵐ	−43° 25'	5.7, 5.7	0.8"	A5	Binary – period 190 yrs.
Σ2289 Her	18ʰ 10ᵐ	+16° 29'	6.5, 7.2	1.2"	B9V, F2	Very slow binary.
73 Oph	18ʰ 10ᵐ	+04° 00'	6.1, 7.0	0.6"	F2V	Tight 270-yr. binary.
μ Sgr	18ᵍ 14ᵐ	−21° 04'	3.9, 9.8, 9.3	48", 50"	B8, B3, —	Delicate wide triple.
η Sgr	18ʰ 18ᵐ	−36° 46'	3.2, 7.8	3"	M4	Close unequal pair.
Σ2306 Sct	18ʰ 22ᵐ	−15° 05'	7.9, 8.6, 9.0	10", 10"	F5	Nice equilateral triple.
39 Dra	18ʰ 24ᵐ	+58° 48'	5.0, 8.0, 7.4	4", 89"	A3V, F6V, F8	Easy double – tough triple.
Σ2315 Her	18ʰ 25ᵐ	+27° 23'	6.5, 7.5	0.8"	A2V	Binary – period 775 yrs.
AC11 Ser	18ʰ 25ᵐ	−01° 35'	6.8, 7.0	0.8"	A9III	Binary – period 240 yrs.
21 Sgr	18ʰ 25ᵐ	−20° 32'	4.9, 7.4	1.8"	K0, A0	Close color/mag. contraster.
59 Ser	18ʰ 27ᵐ	+00° 12'	5.3, 7.6	4"	A0	Nice close double.
κ CrA	18ʰ 33ᵐ	−38° 44'	5.6, 6.3	21"	B9V, A0III	Neat bluish, fixed combo.
OΣ359 Her	18ʰ 35ᵐ	+23° 36'	6.3, 6.5	0.7"	G9III	Binary – period 210 yrs.
OΣ358 Her	18ʰ 36ᵐ	+16° 59'	6.8, 7.0	1.3"	G2V, G2V	Matched 290-yr. binary.
λ CrA	18ʰ 44ᵐ	−38° 19'	5.1, 9.7	29"	A0, K0	Color/mag. contrast pair.
5 Aql	18ʰ 46ᵐ	−00° 58'	6.0, 7.8	13"	A0, A0	Nice white pair.

Table A3. Double and Multiple Star Working List *(continued)*

Object/Con	RA	Dec	Mags	Sep	Spec	Remarks
o Dra	18h 51m	+59° 23'	4.8, 7.8	34"	K0	Neat unequal double.
Σ2404 Aql	18h 51m	+10° 59'	6.9, 8.1	4"	K5, K3	Tight orange combo.
OΣ525 Lyr	18h 55m	+33° 58'	6.0, 7.7	45"	G5	Fainter Albireo look-alike.
β648 Lyr	18h 57m	+32° 54'	5.4, 7.5	0.8"	G0V	Tough binary – period 61 yrs.
Σ2438 Dra	18h 58m	+58° 14'	7.1, 7.4	0.9"	A2IV	Matched 260-yr. binary.
11 Aql	18h 59m	+13° 37'	5.2, 8.7	18"	F5	Mag. contrast pair.
ζ Sgr	19h 03m	−29° 53'	3.2, 3.4	0.4"	A2III, A2V	Radiant, tough 21-yr. binary.
γ CrA	19h 06m	−37° 04'	4.8, 5.1	1.3"	F8V, F8V	Yellowish, 120-yr. binary.
Σ2470 Lyr	19h 09m	+34° 58'	7.0, 8.4	14"	B3	Together with Σ2474 forms...
Σ2474 Lyr	19h 09m	+34° 36'	6.8, 8.1	16"	G1, G5	the Double-Double's Double!
Σ2472 Lyr	19h 09m	+37° 55'	7.5, 9	21"	K0	Unequal pair with dim 3rd...
Σ2473 Lyr	19h 09m	+37° 56'	10, 10	6"	F5	star at 75" = Σ2473.
Σ2486 Cyg	19h 12m	+49° 51'	6.6, 6.8	8"	G5, G5	Neat yellowish twin combo.
θ Lyr	19h 16m	+38° 08'	4.4, 9.1, 10.9	100", 100"	K0	Faint, neat equilateral triple.
η Lyr	19h 14m	+39° 09'	4.4, 9.1	28"	B3	Fixed mag. contrast pair.
24 Aql	19h 19m	+00° 20'	6.4, 6.6	423"	K0, F0	Wide binocular combo.
β Sgr	19h 23m	−44° 28'	4.0, 7.1	28"	B8, A3	Nice color/mag. contrast pair.
α+8 Vul	19h 29m	+24° 40'	4.4, 5.8	414"	M, K0	Wide, colorful binocular pair.
Σ2578 Cyg	19h 46m	+36° 05'	6.4, 7.2	15"	A0	Distant 9th-mag. star.
Σ2576 Cyg	19h 46m	+33° 36'	8.3, 8.4	2.7"	K3V	Matched 225-yr. binary.
17 Cyg	19h 46m	+33° 44'	5.0, 9.2, 9.0	26", 135"	F5, K5,–	Delicate triple.
ε Dra	19h 48m	+70° 16'	3.8, 7.4	3"	G8III, F6	Slow tight binary.

Table A3. Double and Multiple Star Working List *(continued)*

Object/Con	RA	Dec	Mags	Sep	Spec	Remarks
ζ Sge	19ʰ 49ᵐ	+19° 09'	5.5, 6.2, 8.7	0.3", 9"	A3V	A-B ultra-close 23-yr. binary.
π Aql	19ʰ 49ᵐ	+11° 49'	6.1, 6.9	1.4"	F2	Fixed tight pair.
χ-1 Cyg	19ʰ 51ᵐ	+32° 55'	4.2–14, 9, 9	26", 135"	S7	Famed variable – dim triple.
α Aql	19ʰ 51ᵐ	+08° 52'	0.8, 9.5, 10	165", 247"	A7V	Altair. Wide optical triple.
ψ Cyg	19ʰ 56ᵐ	+52° 26'	4.9, 7.4	3"	A3	Close mag. contrast pair.
η Cyg	19ʰ 56ᵐ	+35° 05'	3.9, 10, 10	46", 50"	K0	Dim equilateral triple.
OΣ394 Cyg	20ʰ 00ᵐ	+36° 25'	7.1, 9.9	11"	K0	Faintish color contrast duo.
h1470 Cyg	20ʰ 04ᵐ	+38° 20'	7.3, 9.4	29"	M	Dim, red and blue-green pair.
15 Sge	20ʰ 04ᵐ	+17° 04'	5.9, 6.8	204"	G1V, A2	Also two wide 9ᵗʰ-mags.
κ Cep	20ʰ 09ᵐ	+77° 43'	4.4, 8.4	7"	B9	Close unequal fixed pair.
θ Sge	20ʰ 10ᵐ	+20° 55'	6.5, 8.5, 7.4	12", 84"	F5IV, G5, K2	Fixed delicate triple system.
29 Cyg	20ʰ 14ᵐ	+36° 48'	5.0, 6.6	212"	A0, K5	Wide tinted pair.
Σ2671 Cyg	20ʰ 18ᵐ	+55° 24'	6.1, 7.5	4"	A0	Neat cozy pair.
β Cap	20ʰ 21ᵐ	–14° 47'	3.4, 6.2	205"	KOII, AOIII	Wide, bright – orange and blue.
γ Cyg	20ʰ 22ᵐ	+40° 15'	2.2, 9.9, 10.9	41", 1.8"	F8I	Delicate, close faint triple.
κ-2 Sgr	20ʰ 24ᵐ	–42° 25'	6.0, 6.9	0.8"	A3	Slow binary – closing.
o Cap	20ʰ 30ᵐ	–18° 35'	6.1, 6.6	22"	A3V, A7V	Neat, easy matched pair.
48 Cyg	20ʰ 38ᵐ	+31° 34'	6.9, 7.0	181"	A0, F0	Wide equal double.
β Del	20ʰ 38ᵐ	+14° 36'	4.0, 4.9	0.5"	F5IV	Bright, rapid 27-yr. binary.
49 Cyg	20ʰ 41ᵐ	+32° 18'	5.7, 7.8	3"	K0	Fixed tight colorful pair.
52 Cyg	20ʰ 46ᵐ	+30° 43'	4.2, 9.4	6"	K0	Slow binary – in Veil Nebula.
λ Cyg	20ʰ 47ᵐ	+36° 29'	4.9, 6.1	0.9"	B5V	Close 390-yr. binary.

Table A3. Double and Multiple Star Working List *(continued)*

Object/Con	RA	Dec	Mags	Sep	Spec	Remarks
Σ2725 Del	20ʰ 47ᵐ	+16° 07'	7.6, 8.4	6"	K0	In field with γ Del.
4 Aqr	20ʰ 51ᵐ	−05° 38'	6.4, 7.2	0.4"	F5V, F8V	Ultra-tight 147 yr. binary.
Σ2735 Del	20ʰ 56ᵐ	+04° 32'	6.1, 7.6	2"	G0	Challenging tight pair.
ε Equ	20ʰ 59ᵐ	+04° 18'	6.0, 6.3, 7.1	0.7", 11"	F5III, —, G0V	Close pair 101-yr. binary.
HIV113 Cyg	21ʰ 02ᵐ	+39° 31'	6.5, 10.6	18"	K2	Dim but colorful pair.
λ Equ	21ʰ 02ᵐ	+07° 11'	7.4, 7.4	3"	F8	Neat, identical twin suns.
Σ2744 Aqr	21ʰ 03ᵐ	+01° 32'	6.7, 7.2	1.2"	F5V	Cozy 1500-yr. binary.
12 Aqr	21ʰ 04ᵐ	−05° 49'	5.9, 7.3	3"	F5, A3	Close color contrast pair.
γ Equ	21ʰ 10ᵐ	+10° 08'	4.7, 5.9	353"	F0, A2	Bright ultra-wide pair.
τ Cyg	21ʰ 15ᵐ	+38° 03'	3.8, 6.4	0.7"	F0IV	Rapid binary – period 50 yrs.
1 Peg	21ʰ 22ᵐ	+19° 48'	4.1, 8.2	36"	K0	Easy color/mag. contrast pair.
Σ2799 Peg	21ʰ 29ᵐ	+11° 05'	7.5, 7.5	1.6"	F2	Cozy pair – distant 9ᵗʰ-mag.
Σ2819 Cep	21ʰ 40ᵐ	+57° 35'	7.5, 8.5	12"	F5	In field with following triple.
Σ2816 Cep	21ʰ 40ᵐ	+57° 29'	5.6, 7.7, 7.8	12", 20"	O6	Spectacular triple system.
μ Cyg	21ʰ 44ᵐ	+28° 45'	4.8, 6.1	1.2"	F6V, G2V	Bright, close 500-yr. binary.
ε Peg	21ʰ 44ᵐ	+09° 52'	2.4, 8.5	143"	K2I	Enif. h's Pendulum Star.
κ Peg	21ʰ 45ᵐ	+25° 39'	4.7, 5.0, 10.6	0.3", 14"	F5IV	A-B 12-yr. rapid binary.
Σ2840 Cep	21ʰ 52ᵐ	+55° 48'	5.5, 7.3	18"	B6, A1	Striking bluish-green double.
Σ2841 Peg	21ʰ 54ᵐ	+19° 43'	6.4, 7.9	22"	K0	B is an 8.6 and 8.8, 0.2" pair.
Σ2848 Peg	21ʰ 58ᵐ	+05° 56'	7.2, 7.5	11"	A2	Nice subdued matched duo.
η PsA	22ʰ 01ᵐ	−28° 27'	5.8, 6.8	1.7"	B8	Neat tight fixed pair.
29 Aqr	22ʰ 02ᵐ	−16° 58'	7.2, 7.4	4"	A2	Cozy matched combo.

Table A3. Double and Multiple Star Working List *(continued)*

Object/Con	RA	Dec	Mags	Sep	Spec	Remarks
π Peg	22ʰ 10ᵐ	+33° 11'	4.3, 5.6	900"	F5III, G6III	Binocular pair with 27 Peg.
Σ2883 Cep	22ʰ 11ᵐ	+70° 08'	5.6, 7.6	15"	F2	Easy fixed pair.
Σ2893 Cep	22ʰ 12ᵐ	+73° 04'	6.2, 8.3	29"	G5	Unequal fixed pair.
41 Aqr	22ʰ 14ᵐ	−21° 04'	5.6, 7.1	5"	K0III, F2V	Topaz and blue slow binary.
Σ2890 Lac	22ʰ 15ᵐ	+49° 53'	8.5, 8.5, 9.5	9", 73"	A0	Trio in cluster NGC 7243.
μ1/2 Gru	22ʰ 16ᵐ	−41° 21'	4.8, 5.1	900"	G8III, G8III	Naked-eye/binocular pair.
Σ2894 Lac	22ʰ 19ᵐ	+37° 46'	6.1, 8.3	16"	F0	Fixed pair – distant 9ᵗʰ-mag.
Σ2903 Cep	22ʰ 22ᵐ	+66° 42'	6.7, 6.7	4"	F5, A2	Neat snug identical twins.
53 Aqr	22ʰ 27ᵐ	−16° 45'	6.4, 6.6	3"	G1V, G2V	Yellowish equal slow binary.
Kr60 Cep	22ʰ 28ᵐ	+57° 42'	9.8, 11, 10	3", 75"	dM4, dM6	A-B 44-yr. red-dwarf binary.
δ1/2 Gru	22ʰ 29ᵐ	−43° 30'	3.4, 4.1	900"	G7, M4	Bright, tinted naked-eye pair.
37 Peg	22ʰ 30ᵐ	+04° 26'	5.8, 7.1	0.7"	F5IV	Close, tough 140-yr. binary.
β PsA	22ʰ 32ᵐ	−32° 21'	4.4, 7.9	30"	A0	Mag. contrast pair – optical.
Δ241 PsA	22ʰ 37ᵐ	−31° 40'	5.8, 7.6	90"	K0, K0	Nice wide pale orange duo.
69 Aqr	22ʰ 48ᵐ	−14° 03'	5.8, 9.0	24"	B9	Unequal optical pair.
τ Aqr	22ʰ 50ᵐ	−13° 36'	4.0, 8.5	133"	K5	Wide mag. contrast pair.
γ PsA	22ʰ 52ᵐ	−32° 53'	4.5, 8.0	4"	A0	Slow unequal binary.
52 Peg	22ʰ 59ᵐ	+11° 44'	6.1, 7.4	0.7"	F0V	Tight 290-yr. binary.
θ Gru	23ʰ 07ᵐ	−43° 31'	4.5, 7.0	1.1"	F4	Slow binary; distant 8ᵗʰ mag.
π Cep	23ʰ 08ᵐ	+75° 23'	4.6, 6.6	1.2"	G2III	Binary – period 147 yrs.
89 Aqr	23ʰ 10ᵐ	−22° 27'	5.1, 5.9	0.4"	G2	Challenging tight pair.
o Cep	23ʰ 19ᵐ	+68° 07'	4.9, 7.1	2.8"	K0III, F6V	Tinted 800-yr. binary.

Table A3. Double and Multiple Star Working List *(continued)*

Object/Con	RA	Dec	Mags	Sep	Spec	Remarks
4 Cas	$23^h 25^m$	+62° 17'	5.2, 7.7, 9.6	99", 10"	K5	An 8.6-mag. 215" away.
72 Peg	$23^h 34^m$	+31° 20'	5.7, 5.8	0.5"	K3III	Close, twin 240-yr. binary.
104 Aqr	$23^h 42^m$	−17° 49'	4.9, 7.7	120"	G0	Wide, unequal pair.
107 Aqr	$23^h 46^m$	−18° 41'	5.7, 6.7	7"	F0IV, F2III	Neat slow binary – opening.
6 CAS	$23^h 49^m$	+62° 13'	5.5, 8.0	1.5"	A2	Tough fixed unequal pair.
Σ3050 And	$23^h 59^m$	+33° 43'	6.6, 6.6	1.6"	G0V	Snug twin 355-yr. binary.

Appendix 4

Telescope Limiting Magnitude and Resolution

Listed below are limiting magnitude and resolution values for a variety of common-sized (Size in inches) telescopes in use today, ranging from 2- to 14-inches in aperture. (The 2.4-inch entry represents the ubiquitous 60mm refractor, of which there are perhaps more than any other telescope in the world!)

Values for the minimum visual magnitude (Mag) listed here are for single stars and are only very approximate; experienced keen-eyed observers may see as much as a full magnitude fainter under excellent sky conditions. Companions to visual double stars – especially those in close proximity to a bright primary – are typically much more difficult to see than is a star of the same magnitude placed alone in the eyepiece field. Among the many variables involved here are light pollution, sky conditions, optical quality, mirror and lens coatings, eyepiece design, obstructed or unobstructed optical system, color (spectral type) of the star, and even the age of the observer. Given here in increments of aperture are a few representative limiting magnitudes to serve as a general indication of what can be expected to be seen in different sized telescopes.

Table A4. Telescope Limiting Magnitude and Resolution

Size	Mag	Dawes	Rayleigh	Markowitz
2.0	10.30	2.28	2.75	3.00
2.4		1.90	2.29	2.50
3.0	11.2	1.52	1.83	2.00
3.5		1.30	1.57	1.71
4.0	11.8	1.14	1.38	1.50
4.5		1.01	1.22	1.33
5.0		0.91	1.10	1.20
6.0	12.7	0.76	0.92	1.00
7.0		0.65	0.79	0.86
8.0	13.3	0.57	0.69	0.75
10.0	13.8	0.46	0.55	0.60
11.0		0.42	0.50	0.55
12.0		0.38	0.46	0.50
12.5	14.3	0.36	0.44	0.48
13.0		0.35	0.42	0.46
14.0	14.5	0.33	0.39	0.43

Three different values in arc seconds are listed for resolution, which are for two stars of equal brightness and of about the sixth magnitude. These figures differ significantly for brighter, fainter and, especially, unequal pairs. Dawes is that based on Dawes' Limit ($R = 4.56/A$), Rayleigh on the Rayleigh Criterion ($R = 5.5/D$), and Markowitz on Markowitz's Limit ($R = 6.0/D$). Note that in these equations A (for aperture) and D (for diameter) are the same thing. For more on these relationships, see the section on resolution studies in Chapter 6.

Appendix 5

The Measurement of Visual Double Stars

A valuable reference for those seriously thinking about measuring double stars is the late Charles Worley's 1961 reprint *Visual Observing of Double Stars* from his acclaimed *Sky & Telescope* series of the same title. The section entitled "The Measurement of Visual Double Stars", which discusses the use of a filar micrometer, is especially useful. While the currently popular reticle eyepiece micrometer is both less expensive and easier to use than is the filar micrometer, the latter has long been the traditional instrument for such work. Even those observers who are using more modern devices will find a working knowledge of its operation worthwhile.

Unfortunately, this little booklet has been out of print for some time now. Through the kind permission of Sky Publishing Corporation, its measurement section has been excerpted below. Although the data in its table of 97 visual doubles expressly compiled for measurement by Dr. Worley was updated in 1970, it is now largely out of date due to the orbital motion of the pairs. His comments about this list in the final two paragraphs are of interest and have been retained, but the actual table itself has been dropped. Nearly every pair originally contained in it can be found in the compilations in Chapter 7 and Appendix 3 of this current book.

Charles Worley spent the latter half of his long career at the U.S. Naval Observatory in Washington, DC, where he measured double stars with the Observatory's 26-inch and 12-inch refractors (and occasionally with the 61-inch astrometric reflector at its field station in Flagstaff, Arizona). He was one of the most active observers of visual binaries in the world. Always a kind friend and mentor to any amateur expressing an interest in helping measure his beloved binary stars, he was truly an "observational astronomer" in the finest sense of the term. He wrote as follows:

> The amateur who has acquired a filar micrometer has open to him a boundless field of interesting observational work in measuring double stars.
>
> As mentioned [earlier], the telescope should have excellent optics, and a sturdy, accurately aligned equatorial mounting with a clock drive is essential. For systematic double star work, a refractor of 8-inch aperture or larger is a very effective instrument....
>
> The observer must calibrate the micrometer before measures of double stars can be made. First, he finds the north point on the position-angle circle; this zero point has to be determined independently each night before he begins his observations. Second, he must find the number of seconds of arc corresponding to one revolution of the micrometer screw. Once it has been accurately determined, this value remains practically constant. To find the north point of the position-circle, proceed as follows. Using a low magnification, set the telescope on a star near the celestial equator and in [on] the meridian. Let the star trail along the fixed wire [of the micrometer], correcting any deviations by rotating the micrometer box and adjusting the tangent screw. When the star trails accurately, the wire is pointing east–west. Then, 90 degrees added to (or subtracted from) the position-circle reading gives the north point.

As an example, suppose that the circle reading is 98.6 degrees when a star trails along the fixed wire. Then 8.6 degrees is the north point on the circle. Later that night, in measuring the position angle of a double star, the circle reading is, say, 245.4 degrees. The true position angle is therefore 245.4 degrees – 8.6 degrees = 236.8 degrees.

The determination of R, the value of one revolution of the screw, is more complicated. We shall use the method of directly measuring the difference of declination between two stars whose positions are accurately known. R is found by dividing the number of screw revolutions into the known difference in Declination. This method give R with sufficient precision for reducing measures of close double stars, since the accidental and systematic errors of a measured separation are quite large even under the best conditions.

Tabulated below are three pairs of stars conveniently distributed around the sky. The bright pair 27 and 28 Tauri, Atlas and Pleione in the Pleiades, will probably prove best to use. [Indeed, since they and their famed associated cluster are easily located and readily visible at some hour of the night most nights of the year, the two fainter pairs mentioned have been dropped here.] In principle, the star positions should be individually corrected to the date of the observation by taking into account precession, aberration, and proper motion. By omitting these corrections, simplicity is gained with little loss in accuracy. Though the star positions in the table are for the date 1962.0 [here updated to the current standard epoch 2000.0], the *differences* in declination should remained practically unchanged for many years. A further correction for atmospheric refraction should be made, but if the pairs are observed when within 30 degrees of the zenith, the differential refraction is very small and may be safely neglected.

[Following are the data for these two stars: their designation, visual magnitudes, Right Ascension in hours, minutes and seconds of time, Declination in degrees, minutes and seconds of arc, and the difference in declination between the two in seconds of arc.]

Pleione (28 Tauri)	5.1	03h, 49m, 11.2s	+24°, 08′, 13″	301″.0
Atlas (27 Tauri)	3.6	03h, 49m, 09.7s	+24°, 03′, 12″	301″.0

The measurements to determine R should be made when the star pair is near the meridian. First, using a low power, carefully find the zero of the position-angle circle, and clamp the wires in the east–west direction. Then, with a higher power, make repeated measurements of the separation in declination of the star pair, by the technique described below. Measure the pair in the north-to-south direction, then rotate the micrometer 180 degrees and measure from south to north.

If the field of view is too small to encompass both stars simultaneously, one or more intermediate stars may be used. These should lie on nearly the same line as the star pair, but their exact positions are immaterial. The observations of 27 Tauri and 28 Tauri [below] are an example:

N-S: 28 Tau to intermediate star – 10.46 revolutions. Intermediate star to 27 Tau – 14.56 revolutions. Sum = 25.02
S-N: 27 Tau to intermediate star – 14.53 revolutions. Intermediate star to 28 Tau – 10.47 revolutions. Sum = 25.00
Average = 25.01
Then R = 301.0″/25.01 = 12″.04 per revolution of the screw.

Such determinations of R should be repeated on a number of nights, and the mean used in the reduction of double star measures.

We finally come to the problem of making an actual measurement of a double star. Observations should be attempted only on nights of good seeing, and the highest power that the seeing will permit should always be employed.

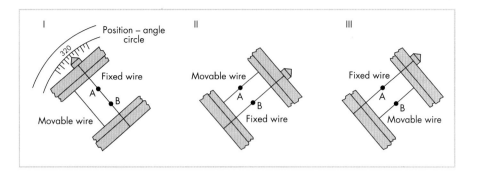

Figure A5.1. The setting for finding the position angle of a double star is shown in panel I. Panels II and III show that required for determining the separation of a pair. Two setting are needed for the separation, the difference between II and III giving twice its actual value as the moveable wire is shifted from one side of the fixed wire to the other. Courtesy *Sky & Telescope*.

The diagram labeled I shows the method of measuring the position angle of the double. With the aid of the tangent screw and position-angle pinion, the fixed wire is set across the star images A and B as accurately as possible. During this operation, the line joining the observer's eyes should always be held either parallel or perpendicular to the line joining the stars. When satisfied that the wire is properly oriented, read the position circle and give the micrometer box an arbitrary turn. This helps insure that successive settings are independent of one another. It is desirable to make four settings for position angle.

To measure the separation, rotate the micrometer box by 90 degrees and clamp it. Two or three settings are made with the wires in the position shown in the diagram that is labeled II, followed by an equal number as shown in diagram III. The interchange of wires eliminates the need to find the coincidence reading of the two wires, and gives the double distance. Since we know R, we find the separation, ρ, of the pair, in seconds of arc, from $\rho = {}^1/_2 R$ *(second reading – first reading)*. Note that separation measures are always made in the order of increasing readings of the screw, in order to eliminate backlash. In making such distance measures, you can lessen bias by taking your eye from the ocular and randomly turning the micrometer screw between settings.

An example of an actual observation made by the author with the Lick Observatory 12-inch refractor is shown. It was made on the double $\Sigma 1932$, on May 27, 1960. The practice of double star observers is to report the date of every measure, dividing the year decimally, 1960.402 in this case. To reduce accidental errors, the double star should be measured on three or more nights in the same season, and an average obtained.

In this article, only a brief outline of techniques can be given. The reader will find much information on micrometers and observing methods in R.G. Aitken's book, *The Binary Stars* (1935). Also worth consulting are J.B. Sidgwick's *Observational Astronomy for Amateurs* (1955) and *Amateur Astronomer's Handbook* (1955), as well as the chapter on micrometers in *Amateur Telescope Making – Book II* (1949). See also an article by W.H. van den Bos, "Some Hints for Double-Star Observers and Orbit Computers," in *Publications* of the Astronomical Society of the Pacific, 70, 160, 1958.

In the following list are 97 interesting double stars suitable for micrometric observations with 6-inch to 12-inch telescopes. Current position angles and separations are listed both as an aid in identification, and to help the casual observer test the resolving power of his telescope.

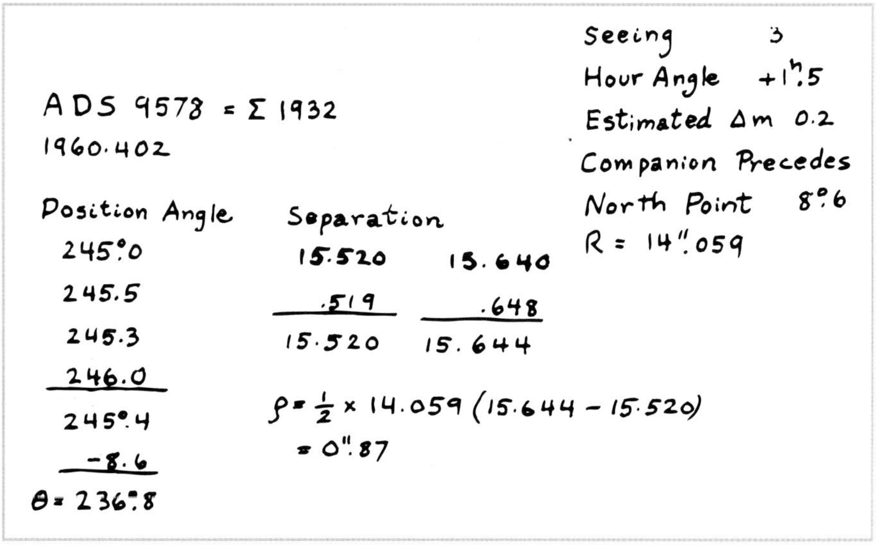

```
ADS 9578 = Σ 1932            Seeing         3
1960.402                      Hour Angle   +1ʰ.5
                              Estimated Δm  0.2
Position Angle   Separation   Companion Precedes
   245°.0         15.520    15.640   North Point   8°.6
   245.5           .519      .648    R = 14".059
   245.3          15.520    15.644
   246.0
   245°.4         ρ = ½ × 14.059 (15.644 − 15.520)
    −8.6            = 0".87
θ = 236°.8
```

Figure A5.2. A sample double star observation recorded on an index card by Charles Worley using the 12-inch refractor of the Lick Observatory on the night of May 27, 1960. Courtesy *Sky & Telescope*.

All of the pairs listed here show appreciable orbital motion. In the last column, the period of revolution is given when known. Uncertain values are marked with a colon (:), very uncertain ones with a double colon (::). Asterisks indicate notes on the last page.

About the Author

James Mullaney is an astronomy writer, lecturer and consultant who has published more than 500 articles and five books on observing the wonders of the heavens, and logged nearly 25,000 hours of stargazing time with the unaided eye, binoculars and telescopes. Formerly Curator of the Buhl Planetarium and Institute of Popular Science in Pittsburgh and more recently Director of the Dupont Planetarium, he served as staff astronomer at the University of Pittsburgh's Allegheny Observatory and as an editor for *Sky & Telescope*, *Astronomy* and *Star & Sky* magazines. One of the contributors to Carl Sagan's award-winning *Cosmos* PBS-Television series, his work has received recognition from such notables (and fellow stargazers) as Sir Arthur Clarke, Johnny Carson, Ray Bradbury, Dr. Wernher von Braun and former student – NASA scientist/astronaut Dr. Jay Apt. His 50-year mission as a "celestial evangelist" has been to "celebrate the universe!" – to get others to look up at the majesty of the night sky and to personally experience the joys of stargazing.

The author, shown holding a copy of his book *Celestial Harvest: 300-Plus Showpieces of the Heavens for Telescope Viewing & Contemplation*. Originally self-published in 1998 (and updated in 2000), it was reprinted in 2002 by Dover Publications in New York. More than 40 years in the making, nearly half of its entries are attractive visual double and multiple stars. Courtesy of *Sky & Telescope* and Warren Greenwald.

Index

With the exception of a number of famous/well-known objects (mainly those having proper names) discussed in one or more places in the text itself, double and multiple stars themselves are not listed in this index since all of the 500 pairs covered in this book can be readily found in either the showpiece roster in **Chapter 7** or in the extended working list in **Appendix 3**. Several of the "first-magnitude" and other single stars are also included here, as they are mentioned in the main text in various contexts relating to their double and multiple star kin.

Aitken, R.G., 7, 67, 69, 70, 100, 125
Albireo (β Cygni), 4, 10, 62, 73, 78, 85
Aldebaran (α Tauri), 105
Algol (β Persei), 18
Allegheny Observatory 30-inch refractor, vii, 44, 53
Algieba (γ Leonis), 80
Almach (γ Andromedae), 4, 62, 76
α Capricorni, 9, 77
α Centauri, 16
AM Canum Venaticorum, 30
American Association of Variable Star Observers (AAVSO), 74, 87
Angular measure, 47
Angular separation, 14, 68
Antares (α Scorpii), 23, 62, 82
Anton, Rainer, 54, 55, 70, 72
Apastron, 12
Aperture masks, 57
Apochromatic refractor, 44
Arcturus (α Bootis), 37
Arguelles-Barbera difficulty index, 64, 65
Argyle, Robert, 70, 74, 86
Asimov, Isaac, 23
Association of Binary Star Observers, 74
Astronomical Unit (AU), 25

Barlow lenses, 49, 70
Barnard/Barnard's Star, 15, 67
Barns, C.E., v, vi
Bernhard, Bennett and Rice, 61
β Lyrae, 19, 81
Betelgeuse (α Orionis), 23, 37, 107
Binaries
—astrometric, 14
—black hole, 21
—cataclysmic, 21
—contact, 19, 28
—eclipsing, 17
—genesis of, 24
—interferometric, 14
—neutron star/pulsar, 21, 22
—spectroscopic, 16
—spectrum/symbiotic, 17
—visual, 12
Biesbroeck, George van, 6, 47
Binoculars, v, 75, 101
Blinking Planetary (NGC 6826), 36
British Astronomical Association (BAA), 74
Brown dwarfs, 15
Burnham, Robert, Jr., 61, 89, 90
Burnham, S.W., 38, 70

Capella (α Aurigae), 17, 106
Cassegrain reflector, 45
Castor (α Geminorum), 5, 11, 12, 17, 79
CCD (charge coupled device) imaging, 53
Celestial Harvest, 61, 131

Center for High Resolution Astronomy (CHRA), 14
Color studies, 61
Computers, 58
Constellations, 91
Cor Caroli (α Canum Venaticorum), 55, 77, 79
Couteau, Paul, 69

W.R. Dawes/Dawes' Limit, 62, 64, 67
Deep-sky objects, 3, 9
Dew caps/light shields, 57
Doppler shift, 16
Double-Double (ε Lyrae), 5, 21, 66, 81
Double stars
—catalogues, 55, 94
—common proper motion, 11
—defined, 3, 9
—demise of field, 6
—discoverers, 94
—distances, 25
—distribution, 23
—extended working list, 101
—frequency, 24
—luminosity, 27
—masses, 25
—optical, 9, 10
—physical, 9
—showpiece list, 75
—sightseeing tour, 61
Double star catalogues and designations, 55, 94
Double star observers' society, founding a, 74
Dwarf novae, 21

Index

Eddington, Sir Arthur, 27
ϵ Aurigae, 19
Erecting prisms, 49
η Orionis, 66, 81
Eye, training
—averted vision, 36
—color perception, 37
—dark adaptation, 37
—visual acuity, 35
Eyepieces
—actual field, 47
—antireflection coatings, 48
—apparent field, 47
—barrel diameters, 48
—binocular, 56
—Erfle, 47
—heaters, 57
—Kellner, 47
—magnification, 48
—Nagler, 47
—Orthoscopic, 47, 63
—Plossl, 47, 63

Fienberg, Rick, 11, 12
Filters, 58
Focal ratio, 43

γ Delphini, 62, 79
Globular clusters, 30
Goodricke, John, 18
Go-To systems, 58
GPS systems, 58

Haas, Sissy, 5, 60, 62, 86
h3780 (NGC 2017), 21, 106
Harrington, Philip, 48
Hartmann, William, 30
Herschel, Sir William, 12, 35, 59, 67, 73
Herschel's Wonder Star (β Monocerotis), 5, 19, 55, 81
Hertzsprung–Russell Diagram, 23, 28
Hipparcos astrometry satellite, 9, 10, 15
Hubble Space Telescope (HST), 30, 46

Index Catalogue of Visual Double Stars (IDS), 94, 100
International Astronomical Union (IAU), 75, 91, 101

Izar (ϵ Bootis), 62, 66, 76

Jones, K.G., 70, 90

Keck Observatory 400-inch binocular telescope, 14, 46
Kepler's laws, 25, 27

Lagrangian lobe/surface, 19, 28
Lightyear (LY), 26, 75, 101
Lick Observatory 36-inch refractor, 44
Limiting magnitude, 61, 121
Longfellow, Henry Wadsworth, 5
Lord (Christopher) Nomogram, 67

MacRobert, Alan, 10
Magnitude
—absolute (intrinsic), 28
—bolometric, 28
—difference/term, 66
—visual (apparent), 28, 66, 75, 101
Main Sequence, 23
Maksutov-Cassegrain systems, 46
Markowitz Limit, 66
Mass exchange, 28
Mass–Luminosity Relation, 27
Maurer, Andreas, 13
McDonald Observatory 82-inch reflector, 47
Meudon Observatory 33-inch refractor, 44
Meridian, 63
Micrometers
—chronometric, 52
—diffraction grating, 52
—double image, 52
—filar, 50, 69, 123
—measurement, 67, 123
—reticle eyepiece, 52
Milky Way Galaxy, 23
Mizar (ζ Ursae Majoris), 5, 9, 17, 83
Mountings and motor (clock) drives, 58
Mount Wilson Observatory 100-inch reflector, 17

Multiple star systems, 19
Muirden, James, 88

Negative observations, value of, 42
New General Catalogue of Double Stars Within 120° of the North Pole (ADS), 100
New pair survey, 73
Newtonian reflector, 44
Newton, Sir Isaac, 25, 27
Norton's Star Atlas, 55
Novae, 21

Observatories, private, 59
Observing guides, classic, 61
Olcott, W.T., 61
Open (galactic) clusters, 23
Orbit calculation/plotting
—dynamical elements, 72
—geometrical elements, 72
—graphical methods, 73
—periods, 73
Orion Nebula, 20, 21, 82

Palomar Observatory 200-inch (Hale) reflector, 46, 89
Parallax
—dynamical, 27
—trigonometric, 25
—spectroscopic, 27
Peltier, Leslie, 87
Periastron, 12
Personal matters
—aesthetic and philosophical considerations, 88
—diet, 42
—dress, 42
—pleasure vs. serious observing, 87
—posture, 42
—rest, 42
Peterson (Harold) diagram, 63, 66
Photographic imaging, 53
Photographer's cloth, 57
Planets (extrasolar), 31
Polaris (α Ursae Minoris), 39, 83
Porrima (γ Virginis), 5, 13, 68, 72, 84

Position angle, 50, 68
Pound, J., 13
Protostars, 24
Purkinje effect, 37

Quotations
—Atamian, George, 90
—Bahcall, John, 88
—Banachiewicz, T., 88
—Bash, Frank, 90
—Berry, Richard, 89, 90
—Burnham, Robert, Jr, 89, 90
—Byrd, Deborah, 89
—Cain, Lee, 88
—Covington, Michael, 90
—Dickinson, Terence, 89
—Dobson, John, 89
—Everett, Edward, 88
—Frost, Robert, 90
—Funk, Ben, 89
—Hladik, Mark, 89
—Houston, Walter Scott, 88, 89
—Jones, R.M., 90
—Levy, David, 88, 89
—Loftus, Graham, 90
—Lorenzin, Tom, 90
—Newton, Jack, 89
—Olcott, W.T., 89
—Raymo, Chet, 89
—Spevak, Jerry, 89
—Webb, T.W., 89
—Weedman, Daniel, 88

Rasalgethi (α Herculis), 5, 12, 23, 62, 79
Rayleigh Criterion, 64
Record keeping, 41
Reporting/sharing observations, 85
Resolution studies/table, 64, 121
Rigel (β Orionis), 24, 37, 81
Ritchey-Chretien reflector, 46
Roche's Limit lobe/surface, 19, 28

Schedler, Johannes, 4
Schmidt-Cassegrain systems, 46
Setting circles, 58
Sidgwick, J.B., 70, 125
Sirius (α Canis Majoris), 15, 38, 77
61 Cygni, 15, 32, 76
Sky & Telescope, vii, 51, 52, 53, 66, 67, 70, 123
Sky Atlas 2000.0/Catalogue, 56, 75, 101
Sky conditions
—aurorae, 63
—light pollution, 4, 61, 71, 121
—moonlight, 4, 61, 71
—seeing, 37, 39
—seeing scales, 38, 41
—transparency, 38, 71
—transparency scales, 41
Speckle interferometry, 14, 53
Spectral types/classes, 75, 101
Star atlases, 55
Star diagonals, 48
Steele, J.D., 61
Stellar mergers, 29
Struves, 67, 76
Supernovae, 21

Tanguay, Ronald, 51, 70, 74, 87
Teague, Thomas, 69, 70
Telescopes
—astigmatism, 40
—catadioptric, 46
—chromatic aberration, 43
—collimation, 39, 40, 66

Telescopes *(continued)*
—cool-down time, 39
—local "seeing", 40
—optical quality, 39, 66, 121, 123
—reflecting, 44
—refracting, 43
—spherical aberration, 40
—star testing (extrafocal image test), 39
Trapezium (θ-1 Orionis), 5, 20, 21, 82

U.S. Naval Observatory (USNO)
—26-inch refractor, 44, 50, 123
—61-inch astrometric reflector, 15
—staff, vii

Variable stars, 17
Vega (α Lyrae), 37, 81
Video imaging, 53

Washington Double Star Catalog (WDS), vii, 56, 67, 75, 94, 101
Webb Society, 69, 70, 74, 85
Webb, T.W., 61, 74, 85, 89
White dwarfs, 21
Worley, Charles, vii, 50, 70, 123
W Ursae Majoris, 19

ξ Ursae Majoris, 5, 73, 83
X-ray binaries, 21

Yerkes Observatory 40-inch refractor, 6, 7, 44, 47

ζ Herculis, 12, 114
ζ Orionis, 66, 82